Lecture Notes
in Computational Science
and Engineering

47

Editors
Timothy J. Barth
Michael Griebel
David E. Keyes
Risto M. Nieminen
Dirk Roose
Tamar Schlick

Artan Boriçi · Andreas Frommer
Bálint Joó · Anthony Kennedy
Brian Pendleton
Editors

QCD and Numerical Analysis III

Proceedings of the Third International Workshop
on Numerical Analysis and Lattice QCD,
Edinburgh, June–July 2003

With 29 Figures and 15 Tables

Editors

Artan Boriçi, Bálint Joó, Anthony Kennedy, Brian Pendleton
University of Edinburgh
School of Physics
Mayfield Road
EH9 3JZ Edinburgh, United Kingdom
email: borici, bj, tony.kennedy, b.pendleton@ph.ed.ac.uk

Andreas Frommer
Bergische Universität Wuppertal
Fachbereich C – Mathematik und Naturwissenschaften
Gaußstr. 20
42097 Wuppertal, Germany
email: frommer@math.uni-wuppertal.de

Library of Congress Control Number: 2005926502

Mathematics Subject Classification (2000): 82B80, 81T80, 65F10, 65F30, 65F15

ISSN 1439-7358
ISBN-10 3-540-21257-4 Springer Berlin Heidelberg New York
ISBN-13 978-3-540-21257-7 Springer Berlin Heidelberg New York

This work is subject to copyright. All rights are reserved, whether the whole or part of the material is concerned, specifically the rights of translation, reprinting, reuse of illustrations, recitation, broadcasting, reproduction on microfilm or in any other way, and storage in data banks. Duplication of this publication or parts thereof is permitted only under the provisions of the German Copyright Law of September 9, 1965, in its current version, and permission for use must always be obtained from Springer. Violations are liable for prosecution under the German Copyright Law.

Springer is a part of Springer Science+Business Media
springeronline.com
© Springer-Verlag Berlin Heidelberg 2005
Printed in The Netherlands

The use of general descriptive names, registered names, trademarks, etc. in this publication does not imply, even in the absence of a specific statement, that such names are exempt from the relevant protective laws and regulations and therefore free for general use.

Typesetting: by the authors using a Springer TEX macro package
Cover design: *design & production*, Heidelberg

Printed on acid-free paper   SPIN: 10966556   41/TechBooks - 5 4 3 2 1 0

Preface

The Third International Workshop on Numerical Analysis and Lattice QCD took place at the University of Edinburgh from June 30th to July 4th, 2003. It continued a sequence which started in 1995 at the University of Kentucky and continued in 1999 with a workshop at the University of Wuppertal. The aim of these workshops is to bring together applied mathematicians and theoretical physicists to stimulate the exchange of ideas between leading experts in the fields of lattice QCD and numerical analysis. Indeed, the last ten years have seen quite a substantial increase in cooperation between the two scientific communities, and particularly so between numerical linear algebra and lattice QCD.

The workshop was organised jointly by the University of Edinburgh and the UK National e-Science Centre. It promoted scientific progress in lattice QCD as an e-Science activity that encourages close collaboration between the core sciences of physics, mathematics, and computer science.

In order to achieve more realistic computations in lattice quantum field theory substantial progress is required in the exploitation of numerical methods. Recently, there has been much progress in the formulation of lattice chiral symmetry satisfying the Ginsparg–Wilson relation. Methods for implementing such chiral fermions efficiently were the principal subject of this meeting, which, in addition, featured several tutorial talks aiming at introducing the important concepts of one field to colleagues from the other. These proceedings reflect this, being organised in three parts: part I contains introductory survey papers, whereas parts II and III contain latest research results in lattice QCD and in computational methods.

Part I starts with a survey paper by Neuberger on lattice chiral symmetry: it reviews the important mathematical properties and concepts, and related numerical challenges. This article is followed by a contribution of Davies and Higham on numerical techniques for evaluating matrix functions, the matrix sign function being the common link between these two first articles. Then, Boriçi reviews the state-of-the-art for computing the fermion determinant with a focus on Krylov methods. He also shows that another version of fermions respecting chiral symmetry, so-called domain wall fermions, is very closely related to overlap fermions when it comes to numerical computations. Finally, Peardon addresses aspects of stochastic processes and molecular dynamics in QCD simulations. In particular, he reviews the Hybrid Monte Carlo method (HMC), the work-horse of lattice QCD simulations.

Part II starts with a contribution by Boriçi on statistical aspects of the computation of the quark determinant: he suggests using order-statistics estimators rather than noisy methods to eliminate bias, and illustrates this with results for the Schwinger model. The paper by de Forcrand and Jahn studies Monte Carlo for $SU(N)$ Young-Mills theories: instead of the usual approach of accumulating $SU(2)$ updates, they perform overrelaxation in full $SU(N)$ and show that this approach is more efficient in practical simulations. In the next article, Follana considers various improvements of classical staggered fermions: for the pion spectrum he shows that undesirable doublers at light quark masses can indeed be avoided by such improvements. Drummond *et al.* also consider improved gauge actions, now using twisted boundary conditions as an infrared regulator: as they show, the resulting two-loop Landau-mean-links accurately describe high-β Monte Carlo simulations. The contribution by Joó is devoted to the study of potential instabilities in Hybrid Monte Carlo simulations: a theoretical study is presented for the simple harmonic oscillator; implications for (light quark) QCD simulations are discussed and illustrated by numerical experiments. The paper by Liu discusses a canonical ensemble approach to finite baryon density algorithms: several stochastic techniques are required there, including a new Hybrid Noisy Monte Carlo algorithm to reduce large fluctuations. The article by Young, Leinweber and Thomas presents finite-range regularized effective field theory as an efficient tool to study the quark mass variation of QCD observables: this includes regularisation schemes and extrapolation methods for the nucleon mass about the chiral limit.

Part III starts with a paper by Ashby, Kennedy and O'Boyle on a new software package implementing Krylov subspace solvers in a modular manner: the main idea is to gain flexibility and portability by separating the generation of the basis from the actual computation of the iterates. The paper by van den Eshof, Sleijpen and van Gijzen analyses Krylov subspace methods in the presence of only inexact matrix vector products; as an important practical consequence, they are able to devise strategies on how to tune the accuracy requirements yielding an overall fastest method, recursive preconditioning being a major ingredient. Fleming addresses data analysis and modelling for data resulting from lattice QCD calculations: he shows that the field might highly profit from elaborate techniques used elsewhere, like Baysian methods, constrained fitting or total least squares. The paper by Arnold *et al.* compares various Krylov subspace methods for different formulations of the overlap operator; a less known method (SUMR), having had no practical applications so far, turns out to be extremely relevant here. In the next article, Kennedy discusses theoretical and computational aspects of the Zolotarev approximation. This is a closed formula L_∞ best rational approximation to the sign function on two intervals left and right of zero, and its efficient matrix evaluation is of crucial importance in simulations of overlap fermions. Finally, Wenger uses a continued fraction expansion of the sign function to show that overlap fermions are intimately related to the 5-dimensional formulation of lattice chiral symmetry: based on this he shows he that equivalence transformations can be used to make the operators involved better conditioned.

We would like to express our gratitude to the authors of the present volume for their effort in writing their contribution. All papers have undergone a strict refereeing process and we would like to extend our thanks to all the referees for their thorough reviewing. AF and ADK gratefully acknowledge support by the Kavli Institute of Theoretical Physics (KITP), Santa Barbara (supported in part by the National Science Foundation under Grant No. PHY99–07949).

The book cover shows a QCD-simulation of quark confinement, a result from a simulation run at Wuppertal University. We are very thankful to Thomas Lippert and Klaus Schilling for providing us with the picture. Special thanks are also due to the LNCSE series editors and to Thanh-Ha Le Thi from Springer for the always very pleasant and efficient cooperation during the preparation of this volume.

Edinburgh, Santa Barbara, Pahrump, Tirana
March 2005

Artan Boriçi
Andreas Frommer
Bálint Joó
Anthony D. Kennedy
Brian Pendleton

Contents

Part I Surveys

An Introduction to Lattice Chiral Fermions
Herbert Neuberger 3

Computing $f(A)b$ for Matrix Functions f
Philip I. Davies, Nicholas J. Higham 15

Computational Methods for the Fermion Determinant and the Link Between Overlap and Domain Wall Fermions
Artan Boriçi 25

Monte Carlo Simulations of Lattice QCD
Mike Peardon 41

Part II Lattice QCD

Determinant and Order Statistics
Artan Boriçi 57

Monte Carlo Overrelaxation for $SU(N)$ Gauge Theories
Philippe de Forcrand, Oliver Jahn 67

Improved Staggered Fermions
Eduardo Follana 75

Perturbative Landau Gauge Mean Link Tadpole Improvement Factors
I.T. Drummond, A. Hart, R.R. Horgan, L.C. Storoni 83

Reversibility and Instabilities in Hybrid Monte Carlo Simulations
Bálint Joó 91

A Finite Baryon Density Algorithm
Keh-Fei Liu 101

The Nucleon Mass in Chiral Effective Field Theory
Ross D. Young, Derek B. Leinweber, Anthony W. Thomas 113

Part III Computational Methods

A Modular Iterative Solver Package in a Categorical Language
T.J. Ashby, A.D. Kennedy, M.F.P. O'Boyle 123

Iterative Linear System Solvers with Approximate Matrix-vector Products
Jasper van den Eshof, Gerard L.G. Sleijpen, Martin B. van Gijzen 133

What Can Lattice QCD Theorists Learn from NMR Spectroscopists?
George T. Fleming 143

Numerical Methods for the QCD Overlap Operator: II. Optimal Krylov Subspace Methods
Guido Arnold, Nigel Cundy, Jasper van den Eshof, Andreas Frommer, Stefan Krieg, Thomas Lippert, Katrin Schäfer 153

Fast Evaluation of Zolotarev Coefficients
A. D. Kennedy 169

The Overlap Dirac Operator as a Continued Fraction
Urs Wenger 191

Index 199

List of Contributors

G. Arnold
Fachbereich Mathematik und
Naturwissenschaften, Universität
Wuppertal, D-42097 Wuppertal,
Germany
arnold@theorie.physik.uni-
wuppertal.de

T.J. Ashby
School of Physics, University of
Edinburgh, James Clerk Maxwell
Building, Mayfield Road, Edinburgh
EH9 3JZ, United Kingdom
t.ashby@ed.ac.uk

A. Boriçi
School of Physics, University of
Edinburgh, James Clerk Maxwell
Building, Mayfield Road, Edinburgh
EH9 3JZ, United Kingdom
borici@ph.ed.ac.uk

N. Cundy
Fachbereich Mathematik und
Naturwissenschaften, Universität
Wuppertal, D-42097 Wuppertal,
Germany
cundy@theorie.physik.uni-
wuppertal.de

P.I. Davies
School of Mathematics, University of
Manchester, Manchester, M13 9PL,
England
pdavies@ma.man.ac.uk

Ph. de Forcrand
Institute for Theoretical Physics,
ETH Zürich, CH-8093 Zürich,
Switzerland
forcrand@phys.ethz.ch

I.T. Drummond
DAMTP, Cambridge University,
Wilberforce Road, Cambridge
CB3 0WA, United Kingdom
it.drummond@damtp.cam.ac.uk

G. T. Fleming
Jefferson Lab, 12000 Jefferson Ave,
Newport News, VA 23606, USA
flemingg@jlab.org

Eduardo Follana
Department of Physics and
Astronomy, University of Glasgow,
G12 8QQ Glasgow, United Kingdom
e.follana@physics.gla.ac.uk

List of Contributors

A. Frommer
Fachbereich Mathematik und
Naturwissenschaften, Universität
Wuppertal, D-42097 Wuppertal,
Germany
frommer@math.uni-wuppertal.de

A. Hart
School of Physics, University
of Edinburgh, King's Buildings,
Edinburgh EH9 3JZ, United
Kingdom
a.hart@ed.ac.uk

N.J. Higham
School of Mathematics, University of
Manchester, Manchester, M13 9PL,
England
higham@ma.man.ac.uk

R.R. Horgan
DAMTP, Cambridge University,
Wilberforce Road, Cambridge
CB3 0WA, United Kingdom
r.r.horgan@damtp.cam.ac.uk

O. Jahn
Department of Physics, CERN,
Theory Division, CH-1211
Geneva 23, Switzerland
and
LNS, MIT, Cambridge MA 02139,
USA
jahn@mit.edu

Bálint Joó
School of Physics, University of
Edinburgh, James Clerk Maxwell
Building, Mayfield Road, Edinburgh
EH9 3JZ, United Kingdom
bj@ph.ed.ac.uk

A.D. Kennedy
School of Physics, University of
Edinburgh, James Clerk Maxwell
Building, Mayfield Road, Edinburgh
EH9 3JZ, United Kingdom
adk@ph.ed.ac.uk

S. Krieg
Fachbereich Mathematik und
Naturwissenschaften, Universität
Wuppertal, D-42097 Wuppertal,
Germany
krieg@theorie.physik.uni-
wuppertal.de

D. B. Leinweber
Special Research Centre for the
Subatomic Structure of Matter and
Department of Physics, University
of Adelaide, Adelaide SA 5005,
Australia
dleinweb@physics.adelaide.edu.au

Th. Lippert
Fachbereich Mathematik und
Naturwissenschaften, Universität
Wuppertal, D-42097 Wuppertal,
Germany
lippert@theorie.physik.uni-
wuppertal.de

Keh-Fei Liu
Deptartment of Physics and
Astronomy, University of Kentucky,
Lexington, KY 40506, USA
liu@pa.uky.edu

H. Neuberger
Department of Physics and
Astronomy, Rutgers University,
Piscataway, NJ 08540,
USA
neuberg@physics.rutgers.edu

M.F.P. O'Boyle
Institute of Computer Systems
Architecture, University of
Edinburgh, James Clerk Maxwell
Building, Mayfield Road, Edinburgh
EH9 3JZ, United Kingdom
mob@inf.ed.ac.uk

M. Peardon
School of Mathematics, Trinity
College, Dublin 2, Ireland
mjp@maths.tcd.ie

K. Schäfer
Fachbereich Mathematik und
Naturwissenschaften, Universität
Wuppertal, D-42097 Wuppertal,
Germany
schaefer@math.uni-wuppertal.de

G. L.G. Sleijpen
Department of Mathematics, Utrecht
University, P.O. Box 80.010, NL-3508
TA Utrecht, The Netherlands
sleijpen@math.uu.nl

L.C. Storoni
DAMTP, Cambridge University,
Wilberforce Road, Cambridge
CB3 0WA, United Kingdom
lcs20@cam.ac.uk

A.W. Thomas
Special Research Centre for the
Subatomic Structure of Matter and
Department of Physics, University
of Adelaide, Adelaide SA 5005,
Australia
athomas@physics.adelaide.au

J. van den Eshof
Department of Mathematics,
University of Düsseldorf, D-40224,
Düsseldorf, Germany
eshof@am.uni-duesseldorf.de

M.B. van Guizen
CERFACS, 42 Avenue Gaspard
Coriolis, 31057 Toulouse Cedex 01,
France
vangijzen@cerfacs.fr

U. Wenger
Theoretical Physics, Oxford
University, 1 Keble Road, Oxford
OX1 3NP, United Kingdom
and
NIC/DESY Zeuthen,
Platanenallee 6, D–15738 Zeuthen,
Germany
urs.wenger@desy.de

R. D. Young
Special Research Centre for the
Subatomic Structure of Matter and
Department of Physics, University
of Adelaide, Adelaide SA 5005,
Australia
ryoung@physics.adelaide.au

Part I

Surveys

An Introduction to Lattice Chiral Fermions

Herbert Neuberger

Department of Physics and Astronomy, Rutgers University, Piscataway, NJ08540, USA neuberg@physics.rutgers.edu

Summary. This write-up starts by introducing lattice chirality to people possessing a fairly modern mathematical background, but little prior knowledge about modern physics. I then proceed to present two new and speculative ideas.

1 Review

1.1 What are Dirac/Weyl Fermions?

One can think about (Euclidean) Field Theory as of an attempt to define integrals over function spaces [1]. The functions are of different types and are called fields. The integrands consist of a common exponential factor multiplied by various monomials in the fields. The exponential factor is written as $\exp(S)$ where the action S is a functional of the fields. Further restrictions on S are: (1) locality (2) symmetries. Locality means that S can be written as an integral over the base space (space-time) which is the common domain of all fields and the integrand at a point depends at most exponentially weakly on fields at other, remote, space-time points. S is required to be invariant under an all important group of symmetries that act on the fields. In a sense, S is the simplest possible functional obeying the symmetries and generically represents an entire class of more complicated functionals, which are equivalently appropriate for describing the same physics.

Dirac/Weyl fields have two main characteristics: (1) They are Grassmann valued, which means they are anti-commuting objects and (2) there is a form of S, possibly obtained by adding more fields, where the Dirac/Weyl fields, ψ, enter only quadratically. The Grassmann nature of ψ implies that the familiar concept of integration needs to be extended. The definition of integration over Grassmann valued fields is algebraic and for an S where the ψ fields enter quadratically, as in $S = \bar{\psi} K \psi + ...$, requires only the propagator, K^{-1}, and the determinant, $\det K$. Hence, only the linear properties of the operator K come into play, and concepts like a "Grassmann integration measure" are, strictly speaking, meaningless, although they make sense for ordinary, commuting, field integration variables.

Let us focus on a space-time that is a a 4D Euclidean flat four torus, with coordinates x_μ, $\mu = 1, 2, 3, 4$. Introduce the quaternionic basis σ_μ represented by 2×2 matrices:

$$\sigma_1 = \begin{pmatrix} 0 & 1 \\ 1 & 0 \end{pmatrix} \quad \sigma_2 = \begin{pmatrix} 0 & -i \\ i & 0 \end{pmatrix} \quad \sigma_3 = \begin{pmatrix} -1 & 0 \\ 0 & 1 \end{pmatrix} \quad \sigma_4 = \begin{pmatrix} i & 0 \\ 0 & i \end{pmatrix}$$

The ψ fields are split into two kinds, $\bar{\psi}$ and ψ, each being a two component function on the torus. In the absence of other fields the Weyl operators playing the role of the kernel K are $W = \sigma_\mu \partial_\mu$ and $W^\dagger = -\sigma_\mu^\dagger \partial_\mu$. The Dirac operator is made by combining the Weyl operators:

$$D = \begin{pmatrix} 0 & W \\ -W^\dagger & 0 \end{pmatrix} = \begin{pmatrix} 0 & \sigma_\mu \\ \sigma_\mu^\dagger & 0 \end{pmatrix} \partial_\mu \equiv \gamma_\mu \partial_\mu = -D^\dagger$$

The σ_μ obey

$$\sigma_\mu^\dagger \sigma_\nu + \sigma_\nu^\dagger \sigma_\mu = 2\delta_{\mu\nu} \quad \sigma_\mu \sigma_\nu^\dagger + \sigma_\nu \sigma_\mu^\dagger = 2\delta_{\mu\nu}$$

which implies $W^\dagger W = -\partial_\mu \partial_\mu = -\partial^2 \begin{pmatrix} 1 & 0 \\ 0 & 1 \end{pmatrix}$. Thus, one can think about W as a complex square root of the Laplacian. Similarly, one has $D^\dagger D = DD^\dagger = -D^2$, with D^2 being $-\partial^2$ times a 4×4 unit matrix.

When we deal with gauge theories there are other important fields [2]. These are the gauge fields, which define a Lie algebra valued one-form on the torus, denoted by $A \equiv A_\mu dx_\mu$. We shall take $A_\mu(x)$ to be an anti-hermitian, traceless, $N \times N$ matrix. The 1-form defines parallel transport of N-component complex fields Φ by:

$$\Phi(x(1)) = \mathcal{P} e^{\int_\mathcal{C} A \cdot dx} \Phi(x(0))$$

where $x_\mu(t), t \in [0,1]$ is a curve $\mathcal{C}$ connecting $x(0)$ to $x(1)$ and $\mathcal{P}$ denotes path ordering, the ordered product of $N \times N$ matrices being implicit in the exponential symbol. Covariant derivatives, $D_\mu = \partial_\mu - A_\mu$, have as main property the transformation rule:

$$g^\dagger(x) D_\mu(A) g(x) = D_\mu(A^g) \quad A^g \equiv A - g^\dagger dg$$

where the $g(x)$ are unitary $N \times N$ matrices with unit determinant. The replacement of ∂_μ by D_μ is known as the principle of minimal substitution and defines A-dependent Weyl and Dirac operators. A major role is played by local gauge transformations, defined by $\psi \to g\psi, \bar{\psi} \to \bar{\psi} g^\dagger$ and $A \to A^g$ where ψ is viewed as a column and $\bar{\psi}$ as a row. The gauge transformations make up an infinite invariance group and only objects that are invariant under this group are of physical interest. In particular, S itself must be gauge invariant and the ψ dependent part of it is of the form $S_\psi = \int_x \bar{\psi} W \psi$ with W possibly replaced by $W^\dagger$ or by D.

Formally, W^{-1} is gauge covariant and $\det W$ is gauge invariant. Both the construction of W and of D meet with some problems: (1) W may have exact "zero modes", reflecting a nontrivial analytical index. The latter is an integer defined as $\dim \text{Ker} W^\dagger(A) - \dim \text{Ker} W(A)$. It is possible for this integer to be non-zero because the form A is required to be smooth only up to gauge transformations. The space of all A's then splits into a denumerable collection of disconnected components, uniquely labeled by the index. The integration over A is split into a sum over components with associated integrals restricted to each component. (2) $\det W$ cannot always be defined in a gauge invariant way, but $\det(W^\dagger W) = |\det W|^2$ can. Thus, $\det W$ is to be viewed as a certain square root of $|\det W|^2$, but, instead of being a function over the spaces of A it is a line bundle. As a line bundle it can be also viewed as a line bundle over the space of gauge orbits of A, where a single orbit is the collection of all elements A^g for a fixed A and all g. The latter bundle may be twisted, and defy attempts to find a smooth gauge invariant section. When this happens we have an anomaly.

1.2 Why is There a Problem on the Lattice?

Lattice field theory [3] tries to construct the desired functional integral by first replacing space-time by a finite, uniform, toroidal square lattice and subsequently constructing a limit in which the lattice spacing, a, is taken to zero. Before the limit is taken functional integration is replaced by ordinary integration producing well defined quantities. One tries to preserve as much as possible of the desired symmetry, and, in particular, there is a symmetry group of lattice gauge transformations given by $\prod_x SU(N)$, where x denotes now a discrete lattice site.

The one-form A is replaced by a collection of elementary parallel transporters, the link matrices $U_\mu(x)$, which are unitary and effect parallel transport from the site x to the neighboring site to x in the positive μ direction. Traversal in the opposite direction goes with $U_\mu^\dagger(x)$. The fields $\bar\psi$ and ψ are now defined at lattice sites only. As a result, $W, W^\dagger$ become finite square matrices. Here are the main problems faced by this construction: (1) The space of link variables is connected in an obvious way and therefore the index of W will vanish always. Indeed, W is just a square matrix. (2) $\det W$ is always gauge invariant, implying that anomalies are excluded. In particular, there no longer is any need to stop the construction at the intermediate step of a line bundle. These properties show that no matter how we proceed, the limit where the lattice spacing a goes to zero will not have the required flexibility.

1.3 The Basic Idea of the Resolution

The basic idea of the resolution [4] is to reintroduce a certain amount of indeterminacy by adding to the lattice version a new infinite dimensional space in which ψ is an infinite vector, in addition to its other indices. Other fields do not see this space, and different components of ψ are accordingly referred to as flavors. Among all fields, only the ψ fields come in varying flavors. W shall be replaced by a linear operator that acts nontrivially in the new flavor space in addition to its previous actions. The infinite dimensional structure is chosen as simple as possible to provide for, simultaneously, good mathematical control, the emergence of a non-zero index and the necessity of introducing an intermediary construction of $\det W$ as a line bundle [5].

The structure of the lattice W operator is that of a lattice Dirac type operator. This special lattice Dirac operator, D, has a mass, acting linearly in flavor space. With this mass term, the structure of our lattice D is:

$$D = \begin{pmatrix} aM^\dagger & aW \\ -aW^\dagger & aM \end{pmatrix}$$

Only M acts nontrivially in flavor space. To obtain a single Weyl field relevant for the subspace corresponding to small eigenvalues of $-D^2$, the operator M is required to satisfy: (1) the index of M is unity (2) the spectrum of $MM^\dagger$ is bounded from below by a positive number, Λ^2. $(\Lambda a)^2$ is of order unity and kept finite and fixed as $a \to 0$. In practice it is simplest to set the lattice spacing a to unity and take all other quantities to be dimensionless. Dimensional analysis can always be used to restore the powers of a. In the continuum, we always work in the units in which $c = \hbar = 1$. Numerical integration routines never know what a is in length units. The lower bound on $MM^\dagger$ is taken to be of order unity.

The index structure of M ensures that, for eigenvalues of $-D^2$ that are small relative to unity, the relevant space is dominated by vectors with vanishing upper components. These vectors are acted on by the W sub-matrix of D. Moreover, the main contribution comes from the zero mode of M, so, both the infinite flavor space and the extra doubling implicit in using a Dirac operator, become irrelevant for the small eigenvalues of $-D^2$ and their associated eigenspace.

The standard choice for M stems from a paper by Callan and Harvey [6] which has been ported to the lattice by Kaplan [7]. The matrix M is given by a first order differential (or difference) operator of the form $-\partial_s + f(s)$, where s is on the real line and represents flavor space. $f(s)$ is chosen to be the sign function, but could equally well just have different constant absolute values for s positive and for s negative.

The construction of the lattice determinant line bundle will not be reviewed here and we shall skip ahead directly to the overlap Dirac operator.

1.4 The Overlap Dirac Operator

The continuum Dirac operator combines two independent Weyl operators. The Weyl components stay decoupled so long as there is no mass term, and admit independently acting symmetries. Thus, zero mass Dirac fields have more symmetry than massive ones. In particular, this implies that radiative corrections to small Dirac masses must stay proportional to the original mass, to ensure exact vanishing in the higher symmetry case. A major problem in particle physics is to understand why all masses are so much smaller than the energy at which all gauge interactions become of equal strength and one of the most important examples of a possible explanation is provided by the mechanism of chiral symmetry. Until about six years ago it was believed that one could not keep chiral symmetries on the lattice and therefore lattice work with small masses required careful tuning of parameters.

Once we have a way to deal with individual Weyl fermions, it must be possible to combine them pair-wise just as in the continuum and end up with a lattice Dirac operator that is exactly massless by symmetry. This operator is called the overlap Dirac operator and is arrived at by combining the two infinite flavor spaces of each Weyl fermion into a new single infinite space [8]. However, unlike the infinite space associated with each Weyl fermion, the combined space can be viewed as the limit of a finite space. This is so because the Dirac operator does not have an index – unlike the Weyl operator – nor does it have an ill defined determinant. Thus, there is no major problem if the lattice Dirac operator is approximated by a finite matrix. The two flavor spaces are combined simply by running the coordinate s first over the values for one Weyl component and next over the values for the other Weyl component. Since one Weyl component comes as the hermitian conjugate of the other it is no surprise that the coordinate s will be run in opposite direction when it is continued. Thus, one obtains an infinite circle, with a combined function $f(s)$ which is positive on half of the circle and negative on the other. The circle can be made finite and then one has only approximate chiral symmetry [9]. One can analyze the limit when the circle goes to infinity and carry out the needed projection on the small eigenvalue eigenspaces to restrict one to only the components that would survive in the continuum limit. The net result is a formula for the lattice overlap Dirac operator, D_o [8].

To explain this formula one needs, as a first step, to introduce the original lattice Dirac operator due to Wilson, D_W. That matrix is the most sparse one possible with the right symmetry properties, excepting chiral symmetry. It is used as a kernel of the more elaborate construction needed to produce produce D_o. Any alternative to D_W will produce, by the same construction, a new D_o, possibly enhancing some of its other properties. The original D_o is still the most popular, because the numerical advantage of maximal sparseness of D_W has proven hard to beat by benefits coming from other improvements. Thus, we restrict ourselves here only to D_W.

$$D_W = m + 4 - \sum_\mu V_\mu$$
$$V_\mu = \frac{1-\gamma_\mu}{2} T_\mu + \frac{1+\gamma_\mu}{2} T_\mu^\dagger$$
$$\langle x | T_\mu | \Phi^i \rangle = U_\mu(x)^{ij} \langle x | \Phi^j \rangle$$
$$U_\mu(x) U_\mu^\dagger(x) = 1 \quad \gamma_\mu = \begin{pmatrix} 0 & \sigma_\mu \\ \sigma_\mu^\dagger & 0 \end{pmatrix} \quad \gamma_5 = \gamma_1 \gamma_2 \gamma_3 \gamma_4$$

$|\Phi^i\rangle$ is a vector with components labeled by the sites x. The notation indicates that this is the i-th component of a vector $|\Phi\rangle$ with components labeled by both a site x and a group index, j. It is easy to see that $V_\mu V_\mu^\dagger = 1$, so D_W is bounded. $H_W = \gamma_5 D_W$ is hermitian and sparse. The parameter m must be chosen in the interval $(-2, 0)$, and typically is around -1. For gauge fields that are small, the link matrices are close to unity and a sizable interval around zero can be shown to contain no eigenvalues of H_W [10]. This spectral gap can close for certain gauge configurations, but these can be excluded by a simple local condition on the link matrices. When that condition is obeyed, and otherwise independently on the gauge fields, all eigenvalues of H_W^2 are bigger than some positive number μ^2. This makes it possible to unambiguously define the sign function of H_W, $\epsilon(H_W)$. Moreover, $\epsilon(H_W)$ can be infinitely well approximated by a smooth function so long as $\mu^2 > 0$. Since, in addition, the spectrum of H_W is bounded from above, Weierstrass's approximation theorem applies and one can approximate uniformly $\epsilon(H_W)$ by a polynomial in H_W. Thus, as a matrix, ϵ is no longer sparse, but, for $\mu^2 > 0$, it still is true that entries associated with lattice sites separated by distances much larger than $\frac{1}{\mu}$ are exponentially small.

The exclusion of some configurations ruins the simple connectivity of the space of link variables just as needed to provide for a lattice definition of the integer n, which in the continuum labels the different connected components of gauge orbit space. The appropriate definition of n on the lattice is [11]

$$n = \frac{1}{2} Tr \epsilon(H_W)$$

It is obvious that it gives an integer since H_W must have even dimensions as is evident from the structure of the γ-matrices. Moreover, it becomes very clear why configurations for which H_W could have a zero eigenvalue needed to be excised. These configurations were first found to need to be excised when constructing the lattice version of the $\det W$ line bundle.

The overlap Dirac operator is

$$D_o = \frac{1}{2}(1 + \gamma_5 \epsilon(H_W))$$

γ_5 and ϵ make up a so called "Kato pair" with elegant algebraic properties [12].

1.5 What About the Ginsparg-Wilson Relation?

In practice, the inverse of D_o is needed more than D_o itself. Denoting $\gamma_5 \epsilon(H_W) = V$, where V is unitary and obeys "γ_5-hermiticity", $\gamma_5 V \gamma_5 = V^\dagger$, we easily prove that $D_o^{-1} = \frac{2}{1+V}$ obeys

$$\{\gamma_5, D_o^{-1} - 1\} = 0$$

Here, we introduced the anti-commutator $\{a, b\} \equiv ab + ba$. In the continuum, the same relation is obeyed by D^{-1} and reflects chiral symmetry. We see that a slightly altered propagator will be chirally symmetric. The above equation, modifying the continuum relation $\{\gamma_5, D^{-1}\} = 0$, was first written down by Ginsparg and Wilson (GW) in 1982 [14] in a slightly different form. By a quirk of history, their paper became famous only after the discovery of D_o. The main point of the GW paper is that shifting an explicitly chirally symmetric propagator by a matrix which is almost diagonal in lattice sites and unity in spinor space does not destroy physical chiral symmetry.

It turns out that the explicitly chirally symmetric propagator, $\frac{1-V}{1+V}$, can be used as the propagator associated with the monomials of the fields that multiply e^S, but in other places where the propagator appears (loops), one needs to use the more subtly chirally symmetric propagator, $D_o^{-1} = \frac{2}{1+V}$. This dichotomy is well understood and leads to no inconsistencies [15].

Any solution of the GW relation, if combined with γ_5 hermiticity, is of the form $\frac{2}{1+V}$, producing a propagator which anti-commutes with γ_5 of the form $\frac{1-V}{1+V}$. V is a unitary, γ_5-hermitian, matrix. Thus the overlap is the general γ_5-hermitian solution to the GW relation, up to an inconsequential generalization which adds a sparse, positive definite, kernel matrix to the GW relation. The overlap goes beyond the GW paper in providing a generic procedure to produce explicit acceptable matrices V starting from explicit matrices of the same type as H_W.

When the GW relation was first presented, in 1982, the condition of γ_5-hermiticity was not mentioned. The solution was not written in terms of a unitary matrix V, and there was no explicit proposal for the dependence of the solution on the gauge fields. For these reasons, the paper fell into oblivion, until 1997, when D_o was arrived at by a different route. With the benefit of hindsight we see now that it was a mistake not to pursue the GW approach further.

In 1982 neither the mathematical understanding of anomalies - specifically the need to find a natural $U(1)$ bundle replacing the chiral determinant - nor the paramount importance of the index of the Weyl components were fully appreciated. Only after these developments became widely understood did it become possible to approach the problem of lattice chirality from a different angle and be more successful at solving it. The convergence with the original GW insight added a lot of credence to the solution and led to a large number of papers based on the GW relation.

Already in 1982 GW showed that if a solution to their relation were to be found, the slight violation of anti-commutativity with γ_5 that it entailed, indeed was harmless, and even allowed for the correct reproduction of the continuum triangle diagram, the key to calculating anomalies. Thus, there was enough evidence in 1982 that should have motivated people to search harder for a solution, but this did not happen. Rather, the prevailing opinion was that chirality could not be preserved on the lattice. This opinion was fed by a an ongoing research project which attempted to solve the lattice chirality problem by involving extra scalar fields, interacting with the fermions by trilinear (Yukawa) interactions. In this approach one ignored the topological properties of the continuum Dirac operator with respect to the gauge background. The Yukawa models never worked, but the people involved did not attribute this to the failing treatment of topology, and slowly the feeling that chiral symmetry could not be preserved on the lattice took root.

In retrospect, something went wrong in the field's collective thought process, but parallel developments mentioned earlier eventually provided new impetus to deal with the problem correctly. Luckily, this second opportunity was not missed. There was however substantial opposition and even claims that the new approach was not different from the one based on Yukawa interactions, and therefore, was unlikely to be correct [16].

After the discovery of D_o, fifteen years after the GW paper, a flood of new papers, developing the GW approach further, appeared. Because the overlap development already had produced all its new conceptual results by then, no further substantial advance took place. For example, the importance of topology was reaffirmed in a GW framework [17], but the overlap already had completely settled this issue several years earlier. However, this renewed activity generated enough reverberations in the field to finally eradicate the prevailing assumption of the intervening years, that chiral symmetry could not be preserved on the lattice.

1.6 Basic Implementation

Numerically the problem is to evaluate $\epsilon(H_W)$ on a vector, without storing it, basing oneself on the sparseness of H_W. This can be done because, possibly after deflation, the spectrum of H_W has a gap around 0, the point where the sign function is discontinuous. In addition, since H_W is bounded we need to approximate the sign function well only in two disjoint segments, one on the positive real line and the other its mirror image on the negative side. A convenient form is the Higham representation, which introduces $\epsilon_n(x)$ as an approximation to the sign function:

$$\epsilon_n(x) = \begin{cases} \tanh[2n\tanh^{-1}(x)] & \text{for } |x| < 1 \\ \tanh[2n\tanh^{-1}(x^{-1})] & \text{for } |x| > 1 \\ x & \text{for } |x| = 1 \end{cases}$$

Equivalently,

$$\epsilon_n(x) = \frac{(1+x)^{2n} - (1-x)^{2n}}{(1+x)^{2n} + (1-x)^{2n}} = \frac{x}{n}\sum_{s=1}^{n} \frac{1}{x^2\cos^2\left[\frac{\pi}{2n}\left(s-\frac{1}{2}\right)\right] + \sin^2\left[\frac{\pi}{2n}\left(s-\frac{1}{2}\right)\right]}$$

$$\lim_{n\to\infty} \epsilon_n(x) = \text{sign}(x)$$

$\epsilon_n(H_W)\psi$ can be evaluated using a single Conjugate Gradient (CG) iteration with multiple shifts for all the pole terms labeled by s above [18]. The cost in operations is that of a single CG together with an overhead that is linear in n and eventually dominates. The cost in storage is of $2n$ large vectors. The pole representation can be further improved using exact formulae due to Zolotarev who solved the Chebyshev approximation problem analytically for the sign function, thus eliminating the need to use the general algorithm due to Remez. However, for so called quenched simulations, where one replaces $\det D_o$ by unity in the functional integration, the best is to use a double pass [19] version introduced a few years ago but fully understood only recently [20]. In the double pass version storage and number of operations become n-independent for large n, which, for double precision calculations means an n larger than 30 or so. Thus, the precise form of the pole approximation becomes irrelevant and storage requirements are modest. In "embarrassingly parallel" simulations this is the method of choice because it simultaneously attains maximal numerical accuracy and allows maximal exploitation of machine cycles.

When one goes beyond the $\det D_o = 1$ approximation, one needs to reconsider methods that employ order n storage. A discussion of the relevant issues in this case would take us beyond the limits of this presentation; these issues will be covered by other speakers who are true experts.

2 Beyond Overlap/GW?

The overlap merged with GW because both ideas exploited a single real extra coordinate. The starting point of the overlap construction however seems more general, since it would allow a mass matrix in infinite flavor space even if the latter were associated with two or more coordinates. Thus, one asks whether using two extra coordinates might lead to a structurally new construction [21]. While this might not be better in practice, it at least has the potential of producing something different, unattainable if one just sticks to the well understood GW track.

The function $f(s)$ from the overlap is replaced now by two functions $f_1(s_1)$ and $f_2(s_2)$ and the single differential operator $\partial_s + f(s)$ by two such operators, $d_\alpha = \partial_\alpha + f_\alpha(s_\alpha)$. Clearly, d_1 and d_2 commute. A mass matrix with the desired properties can be now constructed as follows:

$$M = \begin{pmatrix} d_1 & -id_2^\dagger \\ id_2 & -d_1^\dagger \end{pmatrix}$$

The two dimensional plane spanned by s_α is split into four quadrants according to the pair of signs of f_α and, formally, the chiral determinant can be written as the trace of four Baxter Corner Transfer Matrices,

$$\text{chiral det} = Tr[K^{\text{I}} K^{\text{II}} K^{\text{III}} K^{\text{IV}}]$$

While this structure is intriguing, I have made no progress yet on understanding whether it provides a natural definition of a $U(1)$ bundle with the right properties. If it does, one could go over to the Dirac case, and an amusing geometrical picture seems to emerge. It is too early to tell whether this idea will lead anywhere or not.

3 Localization and Domain Wall Fermions

3.1 What are Domain Wall Fermions?

Before the form of D_o was derived we had a circular s space with $f(s)$ changing sign at the opposite ends of a diameter. One of the semi-circles can be eliminated by taking $|f(s)|$ to infinity there, leaving us with a half circle that can be straightened into a segment with two approximate Weyl fields localized at its ends. This is known as the domain wall setup, the walls extending into the physical directions of space-time. Keeping the length of the segment finite but large one has approximate chiral symmetry and an operator D_{DW} which acts on many Dirac fields, exactly one of them having a very small effective mass, and the rest having masses of order unity.

The chiral symmetry is only approximate because matrix elements of $\frac{1}{D_{DW}^\dagger D_{DW}}$ connecting entries associated with opposite ends of the segment, L and R, do not vanish exactly. Using a spectral decomposition of $D_{DW}^\dagger D_{DW}$ we have:

$$\langle L|\frac{1}{D_{DW}^\dagger D_{DW}}|R\rangle = \sum_n \frac{1}{\Lambda_n} \langle \Psi_n|R\rangle\langle \Psi_n|L\rangle^* \quad \langle \Psi_n|\Psi_n\rangle = 1$$

Weyl states are localized at L and R and should not connect with each other. So long as the distance between R and L is infinite and $H_W^2 > \mu^2$ this is exactly proven to be the case. For a finite distance S, the correction goes as $e^{-\mu S}$. Unfortunately, μ can be very small numerically and this would require impractically large values of S. Note that the worse situation occurs if one has simultaneously a relatively large wave-function contribution, $|\langle \Psi_n|R\rangle\langle \Psi_n|L\rangle|$, and a small Λ_n. Unfortunately, this worse case seems to come up in practice.

3.2 The Main Practical Problem

As already mentioned, for the purpose of keeping track of $\det D_o$, one may want to keep in the simulation the dependence on the coordinate s, or, what amounts to a logical equivalent, the n fields corresponding to the pole terms in the sign function truncation. This is the main reason to invest resources in domain wall simulations. In my opinion, if one works in the approximation where $\det D_o = 1$ it does not pay to deal with domain wall fermions because it is difficult to safely assess the magnitude of chirality violating effects in different observables.

The main problem faced by practical domain wall simulations is that in the range of interest for strong interaction (QCD) phenomenology H_W, the kernel of the overlap, has eigenstates with very small eigenvalues in absolute value. It turns out that these states are strongly localized in space-time. However, because of approximate translational invariance in s they hybridize into delocalized bands into the extra dimension. As such, they provide channels by which the Weyl modes at the two locations L and R, where the combined $f(s)$ vanishes, communicate with each other, spoiling the chiral symmetry. To boot, these states have small Λ_n. The one way known to eliminate this phenomenon is to take the separation between the Weyl modes to infinity. This leads to the overlap where the problem becomes only of a numerical nature and is manageable by appropriately deflating H_W to avoid the states for which the reconstruction of the sign function by iterative means is too expensive.

3.3 The New Idea

The new idea is to exploit the well known fact that one dimensional random systems typically always localize. The standard approach uses a homogeneous s coordinate;

translations in s would be a symmetry, except for the points at the boundary. Suppose we randomized to some degree the operators H_W strung along the s-direction, randomly breaking s-translations also locally. This would evade hybridization, making localized states in the s direction. Now the hope is that the right amount of disorder would affect only the states made out of the eigenstates of H_W that are localized in the space-time direction because there states have small eigenvalues making the basis for the hybridized states decay slowly in the s direction.

The problem boils down to invent the right kind, and discover the right amount, of randomness that would achieve the above. A simple idea is to randomize somewhat the parameter m in H_W as a function of s. A numerical test of this idea, with a small amount of randomness, has been carried out together with F. Berruto, T. W. Chiu and R. Narayanan. It turned out that the amount of randomness we used was too small to have any sizable effect. The test did show however that if the randomness is very small nothing is lost, so we have something we can smoothly modify away from. However the computational resources needed for a more thorough experiment are beyond our means, so the matter is left unresolved.

4 Final Words

Much progress has been attained on the problem of lattice chirality, both conceptually and in practical implementations. The old wisdom that symmetry is king in field theory has been again proven. However, there is room for further progress and new ideas can still lead to large payoffs; keep your eyes open !

5 Acknowledgments

Many thanks are due to the organizers of the "Third International Workshop on QCD and Numerical Analysis" for an exciting conference. My research is supported in part by the DOE under grant number DE-FG02-01ER41165.

References

1. M. E. Peskin, D. V. Schroeder, An Introduction to Quantum Field Theory, Addison-Wesley, (1995).
2. L. Alvarez-Gaume, Erice School Math. Phys. 0093 (1985).
3. M. Creutz, Quarks, gluons and lattices, Cambridge University Press, (1983).
4. R. Narayanan, H. Neuberger, Phys. Let. B309, 344 (1993).
5. H. Neuberger, Phys. Rev. D59, 085006 (1999); H. Neuberger, Phys. Rev. D63, 014503 (2001); R. Narayanan, H. Neuberger, Nucl. Phys. B443, 305 (1995).
6. C. G. Callan, J. A. Harvey, Nucl. Phys. B250, 427 (1985).
7. D. B. Kaplan, Phys. Lett. B288, 342 (1992).
8. H. Neuberger, Phys. Lett. B417, 141 (1998).
9. H. Neuberger, Phys. Rev. D57, 5417 (1998).
10. H. Neuberger, Phys. Rev. D61, 085015 (2000).

11. R. Narayanan, H. Neuberger, Nucl. Phys. B412, 574 (1994); R. Narayanan, H. Neuberger, Phys. Rev. Lett. 71, 3251 (1993).
12. H. Neuberger, Chin. J. Phys. 38, 533 (2000); T. Kato, Perturbation theory for linear operators, Springer-Verlag, Berlin, (1984).
13. H. Neuberger, Phys. Lett. B427, 353 (1998).
14. P. H. Ginsparg, K. G. Wilson, Phys. Rev. D25, 2649 (1982).
15. H. Neuberger, Nucl. Phys. Proc. Suppl. 73, 697 (1999).
16. M. Golterman, Y. Shamir, Phys. Lett. B353, 84 (1995); K. Jansen, Phys. Rept. 273, 1 (1996).
17. P. Hasenfratz, V. Laliena, F. Niedermayer, Phys. Lett. B427, 125 (1998).
18. H. Neuberger, Phys. Rev. Lett. 81, 4060 (1998).
19. H. Neuberger, Int. J. Mod. Phys. C10, 1051 (1999).
20. Ting-Wai Chiu, Tung-Han Hsieh, Phys. Rev. E68, 066704 (2003).
21. H. Neuberger, hep-lat/0303009 (2003).

Computing $f(A)b$ for Matrix Functions f *

Philip I. Davies and Nicholas J. Higham[†]

School of Mathematics, University of Manchester, Manchester, M13 9PL, England.
pdavies@ma.man.ac.uk, higham@ma.man.ac.uk.

Summary. For matrix functions f we investigate how to compute a matrix-vector product $f(A)b$ without explicitly computing $f(A)$. A general method is described that applies quadrature to the matrix version of the Cauchy integral theorem. Methods specific to the logarithm, based on quadrature, and fractional matrix powers, based on solution of an ordinary differential equation initial value problem, are also presented

1 Introduction

A large literature exists on methods for computing functions $f(A)$ of a square matrix A, ranging from methods for general f to those that exploit properties of particular functions. In this work we consider the problem of computing $y = f(A)b$, for a given matrix A and vector b. Our aim is to develop methods that require less computation than forming $f(A)$ and then multiplying into b.

Motivation for this problem comes from various sources, but particularly from lattice quantum chromodynamics (QCD) computations in chemistry and physics; see [6], [17] and elsewhere in these proceedings. Here, $f(A)b$ must be computed for functions such as $f(A) = A(A^*A)^{-1/2}$, with A very large, sparse, complex and Hermitian. Applications arising in the numerical solution of stochastic differential equations are described in [1], with $f(A) = A^{1/2}$ and A symmetric positive definite. More generally, it might be desired to compute just a single column of $f(A)$, in which case b can be taken to be a unit vector e_i. We mention that Bai, Fahey and Golub [2] treat the problem of computing upper and lower bounds for a quadratic form $u^T f(A)v$, principally for symmetric positive definite A.

We treat general nonsymmetric A and assume that factorization of A is feasible. While our methods are not directly applicable to very large, sparse

---

*This work was supported by Engineering and Physical Sciences Research Council grant GR/R22612.
[†]Supported by a Royal Society-Wolfson Research Merit Award.

A, they should be useful in implementing methods specialized to such A. For example, in the QCD application the Lanczos-based method of [17, Sec. 4.6] requires the computation of $T^{-1/2}e_1$, where T is symmetric tridiagonal, while techniques applying to general f and sparse A and leading to dense subproblems are described by van der Vorst [18].

2 Rational Approximations

A rational approximation $r(A) = q(A)^{-1}p(A) \approx f(A)$, where p and q are polynomials, can be applied directly to the $f(A)b$ problem to give $y = f(A)b \approx r(A)b$ as the solution of $q(A)y = p(A)b$. Forming $q(A)$ is undesirable, so if this formulation is used then iterative methods requiring only matrix-vector products must be used to solve the linear system [18]. It may also be possible to express $r(A)$ in linear partial fraction form, so that y can be computed by solving a sequence of linear systems involving A but not higher powers (an example is given in the next section). The issues here are largely in construction of the approximation $r(A)$, and hence are not particular to the $f(A)b$ problem. See Golub and Van Loan [8, Chap. 11] for a summary of various rational approximation methods.

3 Matrix Logarithm

We consider first the principal logarithm of a matrix $A \in \mathbb{C}^{n \times n}$ with no eigenvalues on $\mathbb{R}^-$ (the closed negative real axis). This logarithm is denoted by $\log A$ and is the unique matrix Y such that $\exp(Y) = A$ and the eigenvalues of Y have imaginary parts lying strictly between $-\pi$ and π. We will exploit the following integral representation, which is given, for example, by Wouk [19].

Theorem 1. *For $A \in \mathbb{C}^{n \times n}$ with no eigenvalues on $\mathbb{R}^-$,*

$$\log(s(A-I)+I) = \int_0^s (A-I)\bigl[t(A-I)+I\bigr]^{-1} dt,$$

and hence

$$\log A = \int_0^1 (A-I)\bigl[t(A-I)+I\bigr]^{-1} dt. \tag{1}$$

Proof. It suffices to prove the result for diagonalizable A [11, Thm. 6.2.27 (2)], and hence it suffices to show that

$$\log(s(x-1)+1) = \int_0^s (x-1)\bigl[t(x-1)+1\bigr]^{-1} dt$$

for $x \in \mathbb{C}$ lying off $\mathbb{R}^-$; this latter equality is immediate. □

The use of quadrature to approximate the integral (1) is investigated by Dieci, Morini and Papini [5]. Quadrature is also directly applicable to our $f(A)b$ problem. We can apply a quadrature rule

$$\int_0^1 g(t)\, dt \approx \sum_{k=1}^m c_k g(t_k) \qquad (2)$$

to (1) to obtain

$$(\log A) b \approx \left(\sum_{k=1}^m c_k \left[t_k (A - I) + I \right]^{-1} \right) (A - I) b. \qquad (3)$$

Unlike when quadrature is used to approximate $\log A$ itself, computational savings accrue from reducing A to a simpler form prior to the evaluation. Since A is a general matrix we compute the Hessenberg reduction

$$A = QHQ^T, \qquad (4)$$

where Q is orthogonal and H is upper Hessenberg, and evaluate

$$(\log A) b \approx Q \sum_{k=1}^m c_k \left[t_k (H - I) + I \right]^{-1} d, \qquad d = Q^T (A - I) b, \qquad (5)$$

where the Hessenberg linear systems are solved by Gaussian elimination with partial pivoting (GEPP). The computation of (4) (with Q maintained in factored form) and the evaluation of (5) cost $(10/3)n^3 + 2mn^2$ flops, whereas evaluation from (3) using GEPP to solve the linear systems costs $(2/3)mn^3$ flops; thus unless $m \lesssim 5$ the Hessenberg reduction approach is the more efficient for large n. If m is so large that $m \gtrsim 32n$ then it is more efficient to employ a (real) Schur decomposition.

Gaussian quadrature is a particularly interesting possibility in (2). It is shown by Dieci, Morini and Papini [5, Thm. 4.3] that applying the m point Gauss-Legendre quadrature rule to (1) produces the rational function $r_m(A - I)$, where $r_m(x)$ is the $[m/m]$ Padé approximant to $\log(1 + x)$, the numerator and denominator of which are polynomials in x of degree m. Padé approximants to $\log(I + X)$ are a powerful tool whose use is explored in detail in [3], [10]. These approximations are normally used only for $\|X\| < 1$, and under this condition Kenney and Laub [13] show that the error in the matrix approximation is bounded by the error in a corresponding scalar approximation:

$$\|r_m(X) - \log(I + X)\| \le |r_m(-\|X\|) - \log(1 - \|X\|)|; \qquad (6)$$

the norm here is any subordinate matrix norm. The well known formula for the error in Gaussian quadrature provides an exact expression for the error $r_m(X) - \log(I + X)$. As shown in [5, Cor. 4.4] this expression can be written as a power series in X when $\|X\| < 1$. Both these approaches provide error bounds for the approximation of $(\log A)\,b$.

We note that in the case where A is symmetric positive definite, the method of Lu [14] for computing $\log(A)$ that uses Padé approximants is readily adapted to compute $(\log A)b$. That method places no restrictions on $\|I - A\|_2$ but it strongly relies on the symmetry and definiteness of A.

In the inverse scaling and squaring method for computing $\log A$ [3], repeated square roots are used to bring A close to I, with subsequent use of the identity $\log A = 2^k \log A^{1/2^k}$. Unfortunately, since each square root requires $O(n^3)$ flops, this approach is not attractive in the context of computing $(\log A)\,b$.

When $\|I - A\| > 1$, we do not have a convenient bound for the error in the m-point Gauss-Legendre approximation to $(\log A)\,b$. While it follows from standard results on the convergence of Gaussian quadrature [4] that the error in our approximation converges to zero as $m \to \infty$, we cannot predict in advance the value of m needed. Therefore for $\|I - A\| > 1$ adaptive quadrature is the most attractive option.

We report numerical experiments for four problems:

$A = \texttt{eye(64)} + 0.5\,\texttt{urandn(64)}, \quad b = \texttt{urandn(64,1)},$
$A = \texttt{eye(64)} + 0.9\,\texttt{urandn(64)}, \quad b = \texttt{urandn(64,1)},$
$A = \texttt{gallery('parter',64)}, \qquad b = \texttt{urandn(64,1)},$
$A = \texttt{gallery('pascal',8)}, \qquad b = \texttt{urandn(8,1)},$

where we have used MATLAB notation. In addition, $\texttt{urandn(m,n)}$ denotes an $m \times n$ matrix formed by first drawing the entries from the normal $N(0,1)$ distribution and then scaling the matrix to have unit 2-norm. The Parter matrix is mildly nonnormal, with eigenvalues lying in the right half-plane on a curve shaped like a "U" rotated anticlockwise through 90 degrees. The Pascal matrix is symmetric positive definite, with eigenvalues ranging in magnitude from 10^{-4} to 10^3.

We computed $y = \log(A)b$ using a modification of the MATLAB adaptive quadrature routine $\texttt{quadl}$, which is based on a 4-point Gauss-Lobatto rule together with a 7-point Kronrod extension [7]. Our modification allows the integration of vector functions. We employ the Hessenberg reduction as in (5). We also computed the m-point Gauss-Legendre approximation, using the smallest m such that the upper bound in (6) was less than the tolerance when $\|I - A\|_2 < 1$, or else by trying $m = 1, 2, 3\ldots$ successively until the absolute error $\|\log(A)b - \widehat{y}\|_2$ was no larger than the tolerance. Three different absolute error tolerances tol were used. The results are reported in Tables 1–4, in which "g evals" denotes the value of m in (5) that $\texttt{quadl}$ effectively uses.

We see from the results that Gauss-Legendre quadrature with an appropriate choice of m is more efficient than the modified $\texttt{quadl}$ in every case. This is not surprising in view of the optimality properties of Gaussian quadrature, and also because adaptive quadrature incurs an overhead in ensuring that an error criterion is satisfied [15]. The inefficiency of adaptive quadrature is particularly notable in Table 1, where at least 18 function evaluations are always required and a much more accurate result than necessary is returned. Recall, however that unless $\|I - A\|_2 > 1$ we have no way of choosing m for the Gauss-Legendre approximation automatically. We also observe that the error bound (6) provides a rather pessimistic bound for the $(\log A)\, b$ error in Table 2.

Table 1. Results for $A =$ eye(64)$+ 0.5\,$urandn(64), $b =$ urandn(64,1). $\|I-A\|_2 = 0.5$, $\|\log(A)\|_2 = 5.3$e-1.

	Adaptive quadrature		Gauss-Legendre		
tol	g evals	Abs. err.	m	Abs. err.	Upper bound in (6)
1e-3	18	1.2e-13	2	7.9e-6	8.4e-4
1e-6	18	1.2e-13	4	1.1e-10	7.6e-7
1e-9	18	1.2e-13	6	4.1e-15	6.7e-10

Table 2. Results for $A =$ eye(64)$+ 0.9\,$urandn(64), $b =$ urandn(64,1). $\|I-A\|_2 = 0.9$, $\|\log(A)\|_2 = 1.0$.

	Adaptive quadrature		Gauss-Legendre		
tol	g evals	Abs. err.	m	Abs. err.	Upper bound in (6)
1e-3	18	5.6e-10	7	9.7e-13	3.2e-4
1e-6	18	5.6e-10	12	3.1e-15	4.7e-7
1e-9	18	5.6e-10	17	3.1e-15	6.8e-10

Table 3. Results for $A =$ gallery('parter',64), $b =$ urandn(64,1). $\|I-A\|_2 = 3.2$, $\|\log(A)\|_2 = 1.9$.

	Adaptive quadrature		Gauss-Legendre	
tol	g evals	Abs. err.	m	Abs. err.
1e-3	48	1.6e-4	8	5.2e-6
1e-6	48	1.4e-10	10	2.5e-7
1e-9	138	1.6e-13	14	5.2e-10

Table 4. Results for $A =$ gallery('pascal',8), $b =$ urandn(64,1). $\|I-A\|_2 = 4.5$e3, $\|\log(A)\|_2 = 8.4$.

	Adaptive quadrature		Gauss-Legendre	
tol	g evals	Abs. err.	m	Abs. err.
1e-3	198	1.1e-5	128	1.0e-3
1e-6	468	2.8e-10	245	1.0e-6
1e-9	1158	5.7e-14	362	9.8e-10

4 Matrix Powers

To compute the action of an arbitrary matrix power on a vector we identify an initial value ODE problem whose solution is the required vector.

Note that for a positive integer p and A having no eigenvalues on $\mathbb{R}^-$, $A^{1/p}$ denotes the principal pth root: the pth root whose eigenvalues lie in the segment $\{z : -\pi/p < \arg(z) < \pi/p\}$. For other fractional α, A^α can be defined as $\exp(\alpha \log A)$.

Theorem 2. *For $A \in \mathbb{C}^{n \times n}$ with no eigenvalues on $\mathbb{R}^-$ and $\alpha \in \mathbb{R}$, the initial value ODE problem.*

$$\frac{dy}{dt} = \alpha(A - I)\big[t(A - I) + I\big]^{-1} y, \qquad y(0) = b, \qquad 0 \le t \le 1, \qquad (7)$$

has a unique solution $y(t) = \big[t(A - I) + I\big]^\alpha b$, and hence $y(1) = A^\alpha b$.

Proof. The existence of a unique solution follows from the fact that the ODE satisfies a Lipschitz condition with Lipschitz constant $\sup_{0 \le t \le 1} \|(A - I)\big[t(A - I) + I\big]^{-1}\| < \infty$. It is easy to check that $y(t)$ is this solution. □

This result is obtained by Allen, Baglama and Boyd [1] in the case $\alpha = 1/2$ and A symmetric positive definite. They propose using an ODE initial value solver to compute $x(1) = A^{1/2}b$.

Applying an ODE solver is the approach we consider here also. The initial value problem can potentially be stiff, depending on α, the matrix A, and the requested accuracy, so some care is needed in choosing a solver. Again, a Hessenberg reduction of A can be used to reduce the cost of evaluating the differential equation.

We report an experiment with the data $A = $ `gallery('parter',64)`, $b = $ `urandn(64,1)`, as used in the previous section, with $\alpha = -1/2$ and $\alpha = 2/5$. We solved the ODE initial value problem with MATLAB's `ode45` function [9, Chap. 12], obtaining the results shown in Table 5; here, tol is the relative error tolerance, and the absolute error tolerance in the function's mixed absolute/relative error test was set to 10^{-3}tol. It is clear from the displayed numbers of successful steps and failed attempts to make a step that `ode45` found the problems relatively easy.

5 General f: Cauchy Integral Theorem

For general f we can represent $y = f(A)b$ using the matrix version of the Cauchy integral theorem:

$$y = \frac{1}{2\pi i} \int_\Gamma f(z)(zI - A)^{-1} b \, dz, \qquad (8)$$

Table 5. Results for $A = $ gallery('parter',64), $b = $ randn(64,1).

$f(A)$	tol	Succ. steps	Fail. atts	ODE evals	Rel. err
$A^{-1/2}$	1e-3	12	0	73	3.5e-8
	1e-6	14	0	85	6.0e-9
	1e-9	40	0	241	7.7e-12
$A^{2/5}$	1e-3	15	0	79	2.8e-8
	1e-6	16	0	91	2.4e-9
	1e-9	54	0	325	1.8e-12

where f is analytic inside a closed contour Γ that encloses the eigenvalues of A. We take for the contour Γ a circle with centre α and radius β,

$$\Gamma: z - \alpha = \beta e^{i\theta}, \quad 0 \leq \theta \leq 2\pi, \tag{9}$$

and then approximate the integral using the repeated trapezium rule. Using $dz = i\beta e^{i\theta} d\theta = id\theta(z(\theta) - \alpha)$, and writing the integrand in (8) as $g(z)$, we obtain

$$\int_\Gamma g(z)dz = i \int_0^{2\pi} (z(\theta) - \alpha)g(z(\theta))\, d\theta. \tag{10}$$

The integral in (10) is a periodic function of θ with period 2π. Applying the n-point repeated trapezium rule to (10) gives

$$\int_\Gamma g(z)\, dz \approx \frac{2\pi i}{n} \sum_{k=0}^{n-1} (z_k - \alpha)g(z_k),$$

where $z_k - \alpha = \beta e^{2\pi k i/n}$, that is, $z_0, \ldots, z_n$ are equally spaced points on the contour Γ (note that since Γ is a circle we have $z_0 = z_n$). When A is real and we take α real it suffices to use just the z_k in the upper half-plane and then take the real part of the result. When applied to periodic functions the repeated trapezium rule can produce far more accurate results than might be expected from the traditional error estimate [4]

$$\int_a^b f(x)dx - T_n(f) = -\frac{(b-a)^3}{12n^2} f''(\xi), \quad a < \xi < b,$$

where $T_n(f)$ denotes the n-point repeated trapezium rule for the function f. The following theorem can be shown using the Euler-Maclaurin formula.

Theorem 3 ([4, p. 137]). *Let $f(x)$ have period 2π and be of class $C^{2k+1}(-\infty, \infty)$ with $|f^{(2k+1)}(x)| \leq M$. Then*

$$\left| \int_0^{2\pi} f(x)\, dx - T_n(f) \right| \leq \frac{4\pi M\, \zeta(2k+1)}{n^{2k+1}},$$

where $\zeta(k) = \sum_{j=1}^\infty j^{-k}$ is the Riemann zeta function.

We want to apply Theorem 3 to the integral (10)

$$\int_0^{2\pi} h(\theta)\,d\theta := \frac{1}{2\pi}\int_0^{2\pi} (z(\theta)-\alpha)f(z(\theta))(z(\theta)I-A)^{-1}b\,d\theta \qquad (11)$$

where $z(\theta) = \alpha + \beta e^{i\theta}$. The integrand is continuously differentiable so we can choose any k in Theorem 3. We need to consider the derivatives of the integrand in (11), which have the form

$$h^{(k)}(\theta) = \frac{i^k}{2\pi}\sum_{j=0}^{k}(z(\theta)-\alpha)^{j+1}\sum_{i=0}^{j}c_{ijk}f^{(j-i)}(z(\theta))(z(\theta)I-A)^{-(1+i)}b, \qquad (12)$$

for certain constants c_{ijk}.

Several terms in (12) can make $|h^{(2k+1)}(\theta)|$ large and therefore make the error bound in Theorem 3 large. First, we have the term $(z(\theta)-\alpha)^{j+1}$ where $0 \le j \le 2k+1$. The term $z(\theta) - \alpha$ has absolute value equal to the radius of the contour, β. Therefore $|h^{(2k+1)}(\theta)|$ is proportional to β^{2k+2} and the error bound for the repeated trapezium rule will be proportional to $\beta(\beta/n)^{2k+1}$. As the contour needs to enclose all the eigenvalues of A, β needs to be large for a matrix with a large spread of eigenvalues. Therefore a large number of points are required to make the error bound small. Second, we have the powers of the resolvent, $(z(\theta)I - A)^{-(1+i)}$, where $0 \le i \le 2k+1$. These powers will have a similar effect to β on $|h^{(2k+1)}(\theta)|$ and therefore on the error bound. These terms can be large if the contour passes too close to the eigenvalues of A; even if the contour keeps well away from the eigenvalues, the terms can be large for a highly nonnormal matrix, as is clear from the theory of pseudospectra [16]. A large resolvent can also make rounding errors in the evaluation of the integrand degrade the computed result, depending on the required accuracy. Third, the derivatives of $f(z)$ can be large: for example, for the square root function near $z = 0$. Finally, the constants c_{ijk} in (12) grow quickly with k. In summary, despite the attractive form of the bound in Theorem 3, rapid decay of the error with n is not guaranteed.

We give two examples to illustrate the performance of the repeated trapezium rule applied to (10). As in the previous sections, we use a Hessenberg reduction of A to reduce the cost of the function evaluations. We consider the computation of $y = A^{1/2}b$. Our first example is generated in MATLAB by

```
A = randn(20)/sqrt(20) + 2*eye(20); b = randn(20,1)
```

We took $\alpha = 2$ and $\beta = 1.4$ in (9), so that the contour does not closely approach any eigenvalue. From Table 6 we can see that the trapezium rule converges rapidly as the number of points increases. This is predicted by the theory since

- β, the radius of the contour, and $\|(z(\theta)I - A)^{-1}\|$ are small,
- Γ does not go near the origin, and therefore $f^{(k)}(z)$ remains of moderate size.

Our second example involves the Pascal matrix of dimensions 4 and 5. The Pascal matrix is symmetric positive definite and has a mix of small and large eigenvalues. As noted above we would like to choose the contour so that it does not go too near the eigenvalues of A and also does not go near the negative real axis, on which the principal square root function is not defined. As a compromise between these conflicting requirements we choose for Γ the circle with centre $(\lambda_{\min} + \lambda_{\max})/2$ and radius $\lambda_{\max}/2$. The results in Table 7 show that the increases in β and $\|(z(\theta)I - A)^{-1}\|$ and the proximity of Γ to the origin cause a big increase in the number of points required for convergence of the repeated trapezium rule. When we repeated the same experiment using the 6×6 Pascal matrix we found that we required over 1 million points to achieve a relative error of 7.0×10^{-5}.

Our conclusion is that the repeated trapezium rule applied to the Cauchy integral formula can be an efficient way to compute $f(A)b$, but the technique is restricted to matrices that are not too nonnormal and whose eigenvalues can be enclosed within a circle of relatively small radius that does not approach singularities of the derivatives of f too closely. We note that Kassam and Trefethen [12] successfully apply the repeated trapezium rule to the Cauchy integral formula to compute certain matrix coefficients in a numerical integration scheme for PDEs, their motivation being accuracy (through avoidance of cancellation) rather than efficiency.

Table 6. Results for A = randn(20)/sqrt(20) + 2*eye(20), b = randn(20,1), $\alpha = 2$, $\beta = 1.4$.

No. points	8	16	32	64	128
Rel. err.	3.0e-2	1.0e-3	1.4e-6	3.0e-12	2.1e-15

Table 7. Results for $A = $ gallery('pascal',n), $b = $ randn(n,1).

$n = 4$, $\alpha = 13.17$, $\beta = 13.15$		$n = 5$, $\alpha = 46.15$, $\beta = 46.14$	
No. points	Rel. err.	No. points	Rel. err.
2^9	1.5e-1	2^{12}	1.6e+0
2^{10}	5.0e-2	2^{13}	6.1e-1
2^{11}	9.4e-3	2^{14}	1.7e-1
2^{12}	4.6e-4	2^{15}	2.2e-2
2^{13}	1.2e-6	2^{16}	4.5e-4
2^{14}	8.9e-12	2^{17}	2.0e-7
2^{14}	3.8e-15	2^{18}	9.9e-14

References

1. E. J. Allen, J. Baglama, and S. K. Boyd. Numerical approximation of the product of the square root of a matrix with a vector. *Linear Algebra Appl.*, 310: 167–181, 2000.
2. Z. Bai, M. Fahey, and G. H. Golub. Some large-scale matrix computation problems. *J. Comput. Appl. Math.*, 74:71–89, 1996.
3. S. H. Cheng, N. J. Higham, C. S. Kenney, and A. J. Laub. Approximating the logarithm of a matrix to specified accuracy. *SIAM J. Matrix Anal. Appl.*, 22 (4):1112–1125, 2001.
4. P. J. Davis and P. Rabinowitz. *Methods of Numerical Integration*. Academic Press, London, second edition, 1984. ISBN 0-12-206360-0. xiv+612 pp.
5. L. Dieci, B. Morini, and A. Papini. Computational techniques for real logarithms of matrices. *SIAM J. Matrix Anal. Appl.*, 17(3):570–593, 1996.
6. R. G. Edwards, U. M. Heller, and R. Narayanan. Chiral fermions on the lattice. *Parallel Comput.*, 25:1395–1407, 1999.
7. W. Gander and W. Gautschi. Adaptive quadrature—revisited. *BIT*, 40(1): 84–101, 2000.
8. G. H. Golub and C. F. Van Loan. *Matrix Computations*. Johns Hopkins University Press, Baltimore, MD, USA, third edition, 1996. ISBN 0-8018-5413-X (hardback), 0-8018-5414-8 (paperback). xxvii+694 pp.
9. D. J. Higham and N. J. Higham. *MATLAB Guide*. Society for Industrial and Applied Mathematics, Philadelphia, PA, USA, second edition, 2005.
10. N. J. Higham. Evaluating Padé approximants of the matrix logarithm. *SIAM J. Matrix Anal. Appl.*, 22(4):1126–1135, 2001.
11. R. A. Horn and C. R. Johnson. *Topics in Matrix Analysis*. Cambridge University Press, 1991. ISBN 0-521-30587-X. viii+607 pp.
12. A.-K. Kassam and L. N. Trefethen. Fourth-order time stepping for stiff PDEs. To appear in SIAM J. Sci. Comput., 2005.
13. C. S. Kenney and A. J. Laub. Padé error estimates for the logarithm of a matrix. *Internat. J. Control*, 50(3):707–730, 1989.
14. Y. Y. Lu. Computing the logarithm of a symmetric positive definite matrix. *Appl. Numer. Math.*, 26:483–496, 1998.
15. J. N. Lyness. When not to use an automatic quadrature routine. *SIAM Rev.*, 25(1):63–87, 1983.
16. L. N. Trefethen and M. Embree. *Spectra and Pseudospectra: The Behavior of Non-Normal Matrices and Operators*. Princeton University Press, Princeton, NJ, USA, 2005.
17. J. van den Eshof, A. Frommer, T. Lippert, K. Schilling, and H. A. van der Vorst. Numerical methods for the QCD overlap operator. I. Sign-function and error bounds. *Computer Physics Communications*, 146:203–224, 2002.
18. H. A. van der Vorst. Solution of $f(A)x = b$ with projection methods for the matrix A. In A. Frommer, T. Lippert, B. Medeke, and K. Schilling, editors, *Numerical Challenges in Lattice Quantum Chromodynamics*, volume 15 of *Lecture Notes in Computational Science and Engineering*, pages 18–28. Springer-Verlag, Berlin, 2000.
19. A. Wouk. Integral representation of the logarithm of matrices and operators. *J. Math. Anal. and Appl.*, 11:131–138, 1965.

Computational Methods for the Fermion Determinant and the Link Between Overlap and Domain Wall Fermions

Artan Boriçi

School of Physics, The University of Edinburgh, James Clerk Maxwell Building, Mayfield Road, Edinburgh EH9 3JZ, borici@ph.ed.ac.uk

Summary. This paper reviews the most popular methods which are used in lattice QCD to compute the determinant of the lattice Dirac operator: Gaussian integral representation and noisy methods. Both of them lead naturally to matrix function problems. We review the most recent development in Krylov subspace evaluation of matrix functions. The second part of the paper reviews the formal relationship and algebraic structure of domain wall and overlap fermions. We review the multigrid algorithm to invert the overlap operator. It is described here as a preconditioned Jacobi iteration where the preconditioner is the Schur complement of a certain block of the truncated overlap matrix.

1 Lattice QCD

Quantum Chromodynamics (QCD) is the quantum theory of interacting quarks and gluons. It should explain the physics of strong force from low to high energies. Due to asymptotic freedom of quarks at high energies, it is possible to carry out perturbative calculations in QCD and thus succeeding in explaining a range of phenomena. At low energies quarks are confined within hadrons and the coupling between them is strong. This requires non-perturbative calculations. The direct approach it is known to be the lattice approach. The lattice regularization of gauge theories was proposed by [1]. It defines the theory in an Euclidean 4-dimensional finite and regular lattice with periodic boundary conditions. Such a theory is known to be Lattice QCD (LQCD). The main task of LQCD is to compute the hadron spectrum and compare it with experiment. But from the beginning it was realised that numerical computation of the LQCD path integral is a daunting task. Hence understanding the nuclear force has ever since become a large-scale computational project.

In introducing the theory we will limit ourselves to the smallest set of definitions that should allow a quick jump into the computational tasks of LQCD.

A fermion field on a regular Euclidean lattice Λ is a Grassmann valued function $\psi_{\mu,c}(x) \in G$, $x = \{x_\mu, \mu = 1, \ldots, 4\} \in \Lambda$ which carries spin and colour indices $\mu = 1, \ldots, 4$, $c = 1, 2, 3$. Grassmann fields are anticommuting fields:

$$\psi_{\mu,c}(x)\psi_{\nu,b}(y) + \psi_{\nu,b}(y)\psi_{\mu,c}(x) = 0$$

for $\psi_{\mu,c}(x), \psi_{\nu,b}(y) \in G$ and $\mu, \nu = 1, \ldots, 4$, $b, c = 1, 2, 3$. In the following we will denote by $\psi(x) \in G_{12}$ the vector field with 12 components corresponding to Grassmann fields of different spin and colour index.

The first and second order differences are defined by the following expressions:

$$\hat{\partial}_\mu \psi(x) = \tfrac{1}{2a}[\psi(x + ae_\mu) - \psi(x - ae_\mu)]$$

$$\hat{\partial}_\mu^2 \psi(x) = \tfrac{1}{a^2}[\psi(x + ae_\mu) + \psi(x - ae_\mu) - 2\psi(x)]$$

where a and e_μ are the lattice spacing and the unit lattice vector along the coordinate $\mu = 1, \ldots, 4$.

Let $U(x)_\mu \in \mathbb{C}^{3\times 3}$ be an unimodular unitary matrix, an element of the $SU(3)$ Lie group in its fundamental representation. It is a map onto $SU(3)$ colour group of the oriented link connecting lattice sites x and $x + ae_\mu$. Physically it represents the gluonic field which mediates the quark interactions represented by the Grassmann fields. A typical quark field interaction on the lattice is given by the bilinear form:

$$\bar{\psi}(x)U(x)_\mu \psi(x + ae_\mu)$$

where $\bar{\psi}(x)$ is a second Grassmann field associated to $x \in \Lambda$. Lattice covariant differences are defined by:

$$\nabla_\mu \psi(x) = \tfrac{1}{2a}[U(x)_\mu \psi(x + ae_\mu) - U^H(x - ae_\mu)_\mu \psi(x - ae_\mu)]$$

$$\Delta_\mu \psi(x) = \tfrac{1}{a^2}[U(x)_\mu \psi(x + ae_\mu) + U^H(x - ae_\mu)_\mu \psi(x - ae_\mu) - 2\psi(x)]$$

where by $U^H(x)$ is denoted the Hermitian conjugation of the gauge field $U(x)$, which acts on the colour components of the Grassmann fields. The Wilson-Dirac operator is a matrix operator $D_W(m_q, U) \in \mathbb{C}^{N\times N}$. It can be defined through 12×12 block matrices $[D_W(m_q, U)](x, y) \in \mathbb{C}^{12\times 12}$ such that:

$$[D_W(m_q, U)\psi^q](x) = m_q \psi^q(x) + \sum_{\mu=1}^{4} [\gamma_\mu \nabla_\mu \psi^q(x) - \frac{a}{2} \Delta_\mu \psi^q(x)]$$

where m_q is the bare quark mass with the index $q = 1, \ldots, N_f$ denoting the quark flavour; $\psi^q(x)$ denotes the Grassmann field corresponding to the quark flavour with mass m_q; $\{\gamma_\mu \in \mathbb{C}^{4\times 4}, \mu = 1, \ldots, 5\}$ is the set of anti-commuting and Hermitian gamma-matrices of the Dirac-Clifford algebra acting on the spin components of the Grassmann fields; $N = 12L_1L_2L_3L_4$ is the total number of fermion fields on a lattice with L_1, L_2, L_3, L_4 sites in each dimension. $D_W(m_q, U)$ is a non-Hermitian operator. The Hermitian Wilson-Dirac operator is defined to be:

$$H_W(m_q, U) = \gamma_5 D_W(m_q, U)$$

where the product by γ_5 should be understood as a product acting on the spin subspace.

The fermion lattice action describing N_f quark flavours is defined by:

$$S_f(U, \psi_1, \ldots, \psi_{N_f}, \bar{\psi}_1, \ldots, \bar{\psi}_{N_f}) = \sum_{q=1}^{N_f} \sum_{x,y \in \Lambda} \bar{\psi}^q(x)[D_W(m_q, U)](x,y)\psi^q(y)$$

The gauge action which describes the dynamics of the gluon field and its interaction to itself is given by:

$$S_g(U) = \frac{1}{g^2} \sum_{\mathbb{P}} \mathrm{Tr}\,(\mathbb{1} - U_{\mathbb{P}})$$

where $\mathbb{P}$ denotes the oriented elementary square on the lattice or the plaquette. The sum in the right hand side is over all plaquettes with both orientations and the trace is over the colour subspace. $U_{\mathbb{P}}$ is a $SU(3)$ matrix defined on the plaquette $\mathbb{P}$ and g is the bare coupling constant of the theory.

The basic computational task in lattice QCD is the evaluation of the path integral:

$$Z_{QCD} = \int \sigma_H(U) \prod_{q=1}^{N_f} \sigma(\psi^q, \bar{\psi}^q) e^{-S_f(U, \psi_1, \ldots, \psi_{N_f}, \bar{\psi}_1, \ldots, \bar{\psi}_{N_f}) - S_g(U)}$$

where $\sigma_H(U)$ and $\sigma(\psi^q, \bar{\psi}^q)$ denote the Haar and Grassmann measures for the qth quark flavour respectively. The Haar measure is a $SU(3)$ group character, whereas the Grassmann measure is defined using the rules of the Berezin integration:

$$\int d\psi_{\mu,c}(x) = 0, \quad \int d\psi_{\mu,c}(x)\psi_{\mu,c}(x) = 1$$

Since the fermionic action is a bilinear form on the Grassmann fields one gets:

$$Z_{QCD} = \int \sigma_H(U) \prod_{q=1}^{N_f} \det D_W(m_q, U) e^{-S_g(U)}$$

Very often we take $N_f = 2$ for two degenerated 'up' and 'down' light quarks, $m_u = m_d$. In general, a path integral has $O(e^N)$ computational complexity which is classified as an NP-hard computing problem [2]. But stochastic estimations of the path integral can be done by $O(N^\alpha)$ complexity with $\alpha \geq 1$. This is indeed the case for the Monte Carlo methods that are used extensively in lattice QCD, a topic which is reviewed in this volume by Mike Peardon [3].

It is clear now that the bottle-neck of any computation in lattice QCD is the complexity of the fermion determinant evaluation. A very often made approximation is to ignore the determinant altogether. Physically this corresponds to a QCD vacuum without quarks, an approximation which gives errors of the order 10% in the computed mass spectrum. This is called the valence or quenched approximation which requires modest computing resources compared to the true theory. To answer the critical question whether QCD is the theory of quarks and gluons it is thus necessary to include the determinant in the path integral evaluation.

Direct methods to compute the determinant of a large and sparse matrix are very expensive and even not adequate for this class of matrices. The complexity of LU decomposition is $O(N^3)$ and it is not feasible for matrices with $N = 1,920,000$ which is the case for a lattice with 20 sites across each dimension. Even $O(N^2)$ methods are still very expensive. Only $O(N)$ methods are feasible for the present computing power for such a large problem.

2 Gaussian Integral Representation: Pseudofermions

The determinant of a positive definite matrix, which can be diagonalised has a Gaussian integral representation. We assume here that we are dealing with a matrix $A \in \mathbb{C}^{N \times N}$ which is Hermitian and positive definite. For example, $A = H_W(m_q, U)^2$. It is easy to show that:

$$\det A = \int \prod_{i=1}^{N} \frac{d\mathrm{Re}(\phi_i) d\mathrm{Im}(\phi_i)}{\pi} \; e^{-\phi^H A^{-1} \phi}$$

The vector field $\phi(x) \in \mathbb{C}^{12}, x \in \Lambda$ that went under the name pseudofermion field [4], has the structure of a fermion field but its components are complex numbers (as opposed to Grassmann numbers for a fermion field).

Pseudofermions have obvious advantages to work with. One can use iterative algorithms to invert A which are well suited for large and sparse problems. The added complexity of an extended variable space of the integrand can be handled easily by Monte Carlo methods.

However, if A is ill-conditioned then any $O(N^\alpha)$ Monte Carlo algorithm, which is used for path integral evaluations is bound to produce small changes in the gauge field. (Of course, an $O(e^N)$ algorithm would allow changes of any size!) Thus, to produce the next statistically independent gauge field one has to perform a large number of matrix inversions which grows proportionally with the condition number. Unfortunately, this is the case in lattice QCD since the unquenching effects in hadron spectrum are expected to come form light quarks, which in turn make the Wilson-Dirac matrix nearly singular.

The situation can be improved if one uses fast inversion algorithms. This was the hope in the early '90 when state of the art solvers were probed and researched for lattice QCD [5, 6]. Although revolutionary for the lattice community of that time, these methods alone could not improve significantly the above picture.

Nonetheless, pseudofermions remain the state of the art representation of the fermion determinant.

3 Noisy Methods

Another approach that was introduced later is the noisy estimation of the fermion determinant [7, 8, 9, 10]. It is based on the identity:

$$\det A = e^{Tr \log A}$$

and the noisy estimation of the trace of the natural logarithm of A.

Let $Z_j \in \{+1, -1\}, j = 1, \ldots, N$ be independent and identically distributed random variables with probabilities:

$$\mathrm{Prob}(Z_j = 1) = \mathrm{Prob}(Z_j = -1) = \frac{1}{2}, \quad j = 1, \ldots, N.$$

Then for the expectation values we get:

$$\mathbb{E}(Z_j) = 0, \quad \mathbb{E}(Z_j Z_k) = \delta_{jk}, \quad j, k = 1, \ldots, N.$$

The following result can be proven without difficulty.

Proposition 1. *Let X be a random variable defined by:*

$$X = Z^T \log A Z, \quad Z^T = (Z_1, Z_2, \ldots, Z_N)$$

Then its expectation μ and variance σ^2 are given by:

$$\mu = \mathbb{E}(X) = Tr \log A, \quad \sigma^2 = \mathbb{E}[(X - \mu)^2] = 2 \sum_{j \neq k} [Re(\log A)_{jk}]^2.$$

To evaluate the matrix logarithm one can use the methods described in [7, 8, 9, 10]. These methods have similar complexity with the inversion algorithms and are subject of the next section.

However, noisy methods give a biased estimation of the determinant. This bias can be reduced by reducing the variance of the estimation. A straightforward way to do this is to take a sample of estimations $X_1, \ldots, X_p$ and to take as estimator their arithmetic mean.

[8] subtract traceless matrices which reduce the error on the determinant from 559% to 17%. [12] proposes a promising control variate technique which can be found in this volume.

Another idea is to suppress or 'freeze' large eigenvalues of the fermion determinant. They are known to be artifacts of a discretised differential operator. This formulation reduces by an order of magnitude unphysical fluctuations induced by lattice gauge fields [11].

A more radical approach is to remove the bias altogether. The idea is to get a noisy estimator of $Tr \log A$ by choosing a certain order statistic $X_{(k)} \in \{X_{(1)} \leq X_{(2)} \ldots \leq X_{(p)}\}$ such that the determinant estimation is unbiased [13]. More on this subject can be found in this volume from the same author [14].

4 Evaluation of Bilinear Forms of Matrix Functions

We describe here a Lanczos method for evaluation of bilinear forms of the type:

$$\mathcal{F}(b, A) = b^T f(A) b \tag{1}$$

where $b \in \mathbb{R}^N$ is a random vector and $f(s)$ is a real and smooth function of $s \in \mathbb{R}_+$.

The Lanczos method described here is similar to the method of [7]. Its viability for lattice QCD computations has been demonstrated in the recent work of [9]. [7] derive their method using quadrature rules and Lanczos polynomials. Here, we give an alternative derivation which is based on the approach of [15, 16, 17]. The Lanczos method enters the derivation as an algorithm for solving linear systems of the form:

$$Ax = b, \quad x \in \mathbb{C}^N . \tag{2}$$

Lanczos algorithm

Algorithm 1 gives n steps of the Lanczos algorithm [18] on the pair (A, b).

The Lanczos vectors $q_1, \ldots, q_n \in \mathbb{C}^N$ can be compactly denoted by the matrix $Q_n = [q_1, \ldots, q_n]$. They are a basis of the Krylov subspace $\mathcal{K}_n = \text{span}\{b, Ab, \ldots, A^{n-1}b\}$. It can be shown that the following identity holds:

Algorithm 1 The Lanczos algorithm

Set $\beta_0 = 0$, $q_0 = o$, $q_1 = b/||b||^2$
for $i = 1, \ldots n$ **do**
    $v = Aq_i$
    $\alpha_i = q_i^\dagger v$
    $v := v - q_i\alpha_i - q_{i-1}\beta_{i-1}$
    $\beta_i = ||v||_2$
    $q_{i+1} = v/\beta_i$
end for

$$AQ_n = Q_n T_n + \beta_n q_{n+1} e_n^T, \qquad q_1 = b/||b||_2 \qquad (3)$$

e_n is the last column of the identity matrix $\mathbb{1}_n \in \mathbb{R}^{n \times n}$ and T_n is the tridiagonal and symmetric matrix given by:

$$T_n = \begin{pmatrix} \alpha_1 & \beta_1 & & \\ \beta_1 & \alpha_2 & \ddots & \\ & \ddots & \ddots & \beta_{n-1} \\ & & \beta_{n-1} & \alpha_n \end{pmatrix} \qquad (4)$$

The matrix (4) is often referred to as the Lanczos matrix. Its eigenvalues, the so called Ritz values, tend to approximate the extreme eigenvalues of the original matrix A as n increases.

To solve the linear system (2) we seek an approximate solution $x_n \in \mathcal{K}_n$ as a linear combination of the Lanczos vectors:

$$x_n = Q_n y_n, \qquad y_n \in \mathbb{C}^n \qquad (5)$$

and project the linear system (2) on to the Krylov subspace $\mathcal{K}_n$:

$$Q_n^\dagger A Q_n y_n = Q_n^\dagger b = Q_n^\dagger q_1 ||b||_2$$

Using (3) and the orthonormality of Lanczos vectors, we obtain:

$$T_n y_n = e_1 ||b||_2$$

where e_1 is the first column of the identity matrix $\mathbb{1}_n$. By substituting y_n into (5) one obtains the approximate solution:

$$x_n = Q_n T_n^{-1} e_1 ||b||_2 \qquad (6)$$

The algorithm of [8] is based on the Padé approximation of the smooth and bounded function $f(.)$ in an interval [19]. Without loss of generality one can assume a diagonal Padé approximation in the interval $s \in (0, 1)$. It can be expressed as a partial fraction expansion. Therefore, one can write:

$$f(s) \approx \sum_{k=1}^m \frac{c_k}{s + d_k}$$

with $c_k \in \mathbb{R}, d_k \geq 0, k = 1, \ldots, m$. Since the approximation error $O(s^{2m+1})$ can be made small enough as m increases, it can be assumed that the right hand side converges to the left hand side as the number of partial fractions becomes large enough. For the bilinear form we obtain:

$$\mathcal{F}(b, A) \approx \sum_{k=1}^{m} b^T \frac{c_k}{A + d_k \mathbb{1}} b \tag{7}$$

Having the partial fraction coefficients one can use a multi-shift iterative solver of [20] to evaluate the right hand side (7). To see how this works, we solve the shifted linear system:

$$(A + d_k \mathbb{1}) x^k = b$$

using the same Krylov subspace $\mathcal{K}_n$. A closer inspection of the Lanczos algorithm, Algorithm 1 suggests that in the presence of the shift d_k we get:

$$\alpha_i^k = \alpha_i + d_k$$

while the rest of the algorithm remains the same. This is the so called shift-invariance of the Lanczos algorithm. From this property and by repeating the same arguments which led to (6) we get:

$$x_n^k = Q_n \frac{1}{T_n + d_k \mathbb{1}_n} e_1 ||b||_2 \tag{8}$$

A Lanczos algorithm for the bilinear form

The algorithm is derived using the Padé approximation of the previous paragraph. First we assume that the linear system (2) is solved to the desired accuracy using the Lanczos algorithm, Algorithm 1 and (6). Using the orthonormality property of the Lanczos vectors and (8) one can show that:

$$\sum_{k=1}^{m} b^T \frac{c_k}{A + d_k \mathbb{1}} b = ||b||^2 \sum_{k=1}^{m} e_1^T \frac{c_k}{T_n + d_k \mathbb{1}_n} e_1 \tag{9}$$

Note however that in presence of roundoff errors the orthogonality of the Lanczos vectors is lost but the result (9) is still valid [9, 21]. For large m the partial fraction sum in the right hand side converges to the matrix function $f(T_n)$. Hence we get:

$$\mathcal{F}(b, A) \approx \hat{\mathcal{F}}_n(b, A) = ||b||^2 e_1^T f(T_n) e_1 \tag{10}$$

Note that the evaluation of the right hand side is a much easier task than the evaluation of the right hand side of (1). A straightforward method is the spectral decomposition of the symmetric and tridiagonal matrix T_n:

$$T_n = Z_n \Omega_n Z_n^T \tag{11}$$

where $\Omega_n \in \mathbb{R}^{n \times n}$ is a diagonal matrix of eigenvalues $\omega_1, \ldots, \omega_n$ of T_n and $Z_n \in \mathbb{R}^{n \times n}$ is the corresponding matrix of eigenvectors, i.e. $Z_n = [z_1, \ldots, z_n]$. From (10) and (11) it is easy to show that (see for example [22]):

$$\hat{\mathcal{F}}_n(b, A) = ||b||^2 e_1^T Z_n f(\Omega_n) Z_n^T e_1 \tag{12}$$

where the function $f(.)$ is now evaluated at individual eigenvalues of the tridiagonal matrix T_n.

The eigenvalues and eigenvectors of a symmetric and tridiagonal matrix can be computed by the QR method with implicit shifts [23]. The method has an $O(n^3)$ complexity. Fortunately, one can compute (12) with only an $O(n^2)$ complexity. Closer inspection of eq. (12) shows that besides the eigenvalues, only the first elements of the eigenvectors are needed:

$$\hat{\mathcal{F}}_n(b, A) = ||b||^2 \sum_{i=1}^{n} z_{1i}^2 f(\omega_i) \qquad (13)$$

It is easy to see that the QR method delivers the eigenvalues and first elements of the eigenvectors with $O(n^2)$ complexity.

A similar formula (13) is suggested by [7]) based on quadrature rules and Lanczos polynomials. The Algorithm 2 is thus another way to compute the bilinear forms of the type (1).

A note on stopping criterion is in order: if $1/\rho_i, i = 1,\ldots$ is the residual error norm of the system $Ax = b$, it is easy to show that:

$$\rho_{i+1}\beta_i + \rho_i\alpha_i + \rho_{i-1}\beta_{i-1} = 0$$

The algorithm is stopped when the linear system $Ax = b$ is solved to ϵ accuracy. This criterion is only a guide and should be carefully considered depending which function we are given. For $f(s) = \sqrt{s}$ this criterion is shown to work in practice [17].

Algorithm 2 The Lanczos algorithm for computing (1)

Set $\beta_0 = 0$, $\rho_1 = 1/||b||_2$, $q_0 = o$, $q_1 = \rho_1 b$
  for $i = 1,\ldots$ do
    $v = Aq_i$
    $\alpha_i = q_i^\dagger v$
    $v := v - q_i\alpha_i - q_{i-1}\beta_{i-1}$
    $\beta_i = ||v||_2$
    $q_{i+1} = v/\beta_i$
    $\rho_{i+1} = -(\rho_i\alpha_i + \rho_{i-1}\beta_{i-1})/\beta_i$
    if $1/|\rho_{i+1}| < \epsilon$ then
      $n = i$
      stop
    end if
  end for
Set $(T_n)_{i,i} = \alpha_i$, $(T_n)_{i+1,i} = (T_n)_{i,i+1} = \beta_i$, otherwise $(T_n)_{i,j} = 0$
Compute ω_i and z_{1i} by the QL method
Evaluate (1) using (13)

The Lanczos algorithm alone has an $O(nN)$ complexity, whereas Algorithm 2 has a greater complexity: $O(nN) + O(n^2)$. For typical applications in lattice QCD the $O(n/N)$ additional relative overhead is small and therefore Algorithm 2 is the recommended algorithm to compute the bilinear form (1).

We stop the iteration when the underlying liner system is solved to the desired accuracy. However, this may be too demanding since the prime interest here is the computation of the bilinear form (1). Therefore, a better stopping criterion is to monitor the convergence of the bilinear form as proposed in [7].

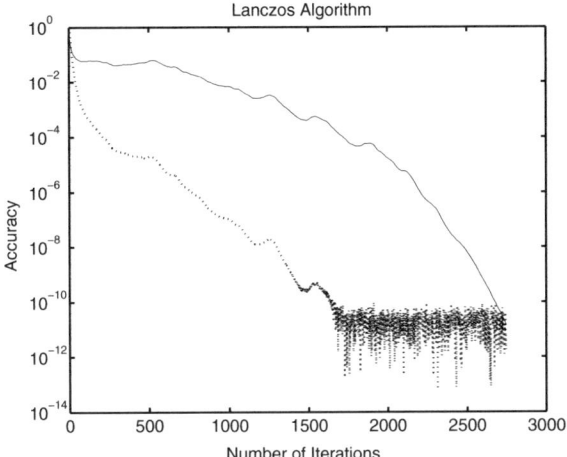

Fig. 1. Normalised recursive residual (solid line) and relative differences of (13) (dotted line) produced by Algorithm 2.

To illustrate this situation we give an example from a $12^3 \times 24$ lattice with $\mu = 0.2$, bare quark mass $m_q = -0.869$ and a $SU(3)$ gauge field background at bare gauge coupling $\beta = 5.9$. We compute the bilinear form (1) for:

$$f(s) = \log \tanh \sqrt{s}, \quad s \in \mathbb{R}_+$$

and $A = H_W(m_q, U)^2$, $b \in \mathbb{R}^N$ and b−elements are chosen randomly from the set $\{+1, -1\}$.

In Fig. 1 are shown the normalised recursive residuals $\|b - Ax_i\|_2 / \|b\|_2$, $i = 1, \ldots, n$ and relative differences of (13) between two successive Lanczos steps. The figure illustrates clearly the different regimes of convergence for the linear system and the bilinear form. The relative differences of the bilinear form converge faster than the computed recursive residual. This example indicates that a stopping criterion based on the solution of the linear system may indeed be strong and demanding. Therefore, the recommended stopping criteria would be to monitor the relative differences of the bilinear form but less frequently than proposed by [7]. More investigations are needed to settle this issue. Note also the roundoff effects (see Fig. 1) in the convergence of the bilinear form which are a manifestation of the finite precision of the machine arithmetic.

5 The Link Between Overlap and Domain Wall Fermions

Wilson regularization of quarks violates chiral symmetry even for massless quarks. This is a serious problem if we would like to compute mass spectrum with light sea quarks. The improvement of the discretization helps to reduce chiral symmetry breaking terms for a Wilson fermion. However, one must go close to the continuum limit in order to benefit from the improvement programme. This is not affordable with the present computing power.

The idea of [24] opened the door for chiral symmetry realization at finite lattice spacing. [25] proposed overlap fermions and [26] domain wall fermions as a theory of chiral fermions on the lattice. The overlap operator is defined by [27]:

$$D(m_q, U) = \frac{1+m_q}{2}\mathbb{1} + \frac{1-m_q}{2}\gamma_5 \text{sign}[H_W(M,U)] \qquad (14)$$

with $M \in (-2, 0)$ which is called the domain wall height, which substitutes the original bare quark mass of Wilson fermions. From now on we suppress the dependence on M, m_q and U of lattice operators for the ease of notations.

Domain wall fermions are lattice fermions in 5-dimensional Euclidean space time similar to the 4-dimensional formulation of Wilson fermions but with special boundary conditions along the fifth dimension. The 5-dimensional domain wall operator can be given by the $L_5 \times L_5$ blocked matrix:

$$\mathcal{M} = \begin{pmatrix} a_5 D_W - \mathbb{1} & P_+ & & -m_q P_- \\ P_- & a_5 D_W - \mathbb{1} & \ddots & \\ & \ddots & \ddots & P_+ \\ -m_q P_+ & & P_- & a_5 D_W - \mathbb{1} \end{pmatrix}, \quad P_\pm = \frac{\mathbb{1}_4 \pm \gamma_5}{2}$$

where the blocks are matrices defined on the 4-dimensional lattices and $P_\pm$ are 4×4 chiral projection operators. Their presence in the blocks with dimensions $N \times N$ should be understood as the direct product with the colour and lattice coordinate spaces. a_5 is the lattice spacing along the fifth dimension.

These two apparently different formulations of chiral fermions are in fact closely related to each other [28] [1]. To see this we must calculate for the domain wall fermions the low energy effective fermion matrix in four dimensions. This can be done by calculating the transfer matrix T along the fifth dimension. Multiplying $\mathcal{M}$ from the right by the permutation matrix:

$$\begin{pmatrix} P_+ & P_- & & \\ & P_+ & \ddots & \\ & & \ddots & P_- \\ P_- & & & P_+ \end{pmatrix}$$

we obtain:

---

[1] For a recent review se also [29]

$$\gamma_5 \begin{pmatrix} (a_5 H_W P_+ - \mathbb{1})(P_+ - m_q P_-) & a_5 H_W P_- + \mathbb{1} & & \\ & a_5 H_W P_+ - \mathbb{1} & \ddots & \\ & & \ddots & a_5 H_W P_- + \mathbb{1} \\ (a_5 H_W P_- + \mathbb{1})(P_- - m_q P_+) & & & a_5 H_W P_+ - \mathbb{1} \end{pmatrix}$$

Further, multiplying this result from the left by the inverse of the diagonal matrix:

$$\begin{pmatrix} a_5 H_W P_+ - \mathbb{1} & & & \\ & a_5 H_W P_+ - \mathbb{1} & & \\ & & \ddots & \\ & & & a_5 H_W P_+ - \mathbb{1} \end{pmatrix}$$

we get:

$$\mathcal{T}(m) := \begin{pmatrix} P_+ - m_q P_- & -T & & \\ & \mathbb{1} & \ddots & \\ & & \ddots & -T \\ -T(P_- - m_q P_+) & & & \mathbb{1} \end{pmatrix}$$

with the transfer matrix T defined by:

$$T = \frac{\mathbb{1}}{\mathbb{1} - a_5 H_W P_+}(\mathbb{1} + a_5 H_W P_-)$$

By requiring the transfer matrix being in the form:

$$T = \frac{\mathbb{1} + a_5 \mathcal{H}_W}{\mathbb{1} - a_5 \mathcal{H}_W}$$

it is easy to see that [28]:

$$\mathcal{H}_W = H_W \frac{1}{2 - a_5 D_W} \quad (15)$$

Finally to derive the four dimensional Dirac operator one has to compute the determinant of the effective fermion theory in four dimensions:

$$\det D^{(L_5)} = \frac{\det \mathcal{T}(m_q)}{\det \mathcal{T}(1)}$$

where the subtraction in the denominator corresponds to a 5-dimensional theory with anti-periodic boundary conditions along the fifth dimension. It is easy to show that the determinant of the $L_5 \times L_5$ block matrix $\mathcal{T}(m_q)$ is given by:

$$\det \mathcal{T}(m_q) = \det[(P_+ - m_q P_-) - T^{L_5}(P_- - m_q P_+)]$$

or

$$\det \mathcal{T}(m_q) = \det[\frac{1 + m_q}{2}\gamma_5(\mathbb{1} + T^{L_5}) + \frac{1 - m_q}{2}(\mathbb{1} - T^{L_5})]$$

which gives:

$$D^{(L_5)} = \frac{1 + m_q}{2}\mathbb{1} + \frac{1 - m_q}{2}\gamma_5 \frac{\mathbb{1} - T^{L_5}}{\mathbb{1} + T^{L_5}} \quad (16)$$

In the large L_5 limit one gets the Neuberger operator (14) but now the operator H_W substituted with the operator $\mathcal{H}_W$ (15). Taking the continuum limit in the fifth dimension one gets $\mathcal{H}_W \to H_W$. This way overlap fermions are a limiting theory of the domain wall fermions. To achieve the chiral properties as in the case of overlap fermions one must take the large L_5 limit in the domain wall formulation.

One can ask the opposite question: is it possible to formulate the overlap in the form of the domain wall fermions? The answer is yes and this is done using truncated overlap fermions [30]. The corresponding domain wall matrix is given by:

$$M_{TOV} = \begin{pmatrix} a_5 D_W - \mathbb{1} & (a_5 D_W + \mathbb{1})P_+ & & -m_q(a_5 D_W + \mathbb{1})P_- \\ (a_5 D_W + \mathbb{1})P_- & a_5 D_W - \mathbb{1} & \ddots & \\ & \ddots & \ddots & (a_5 D_W + \mathbb{1})P_+ \\ -m_q(a_5 D_W + \mathbb{1})P_+ & & (a_5 D_W + \mathbb{1})P_- & a_5 D_W - \mathbb{1} \end{pmatrix}$$

The transfer matrix of truncated overlap fermions is calculated using the same steps as above. One gets:

$$T_{TOV} = \frac{\mathbb{1} + a_5 H_W}{\mathbb{1} - a_5 H_W}$$

The 4-dimensional Dirac operator has the same form as the corresponding operator of the domain wall fermion (16) where T is substituted with T_{TOV} (or $\mathcal{H}_W$ with H_W). Therefore the overlap Dirac operator (14) is recovered in the large L_5 limit.

6 A Two-level Algorithm for Overlap Inversion

In this section we review the two-level algorithm of [31]. The basic structure of the algorithm is that of a preconditioned Jacobi:

$$x^{i+1} = x^i + S_n^{-1}(b - Dx^i), \quad i = 0, 1, \ldots$$

where S_n is the preconditioner of the overlap operator D given by:

$$S_n = \frac{1 + m_q}{2}\mathbb{1} + \frac{1 - m_q}{2}\gamma_5 H_W \sum_{k=1}^{n} \frac{a_k}{H_W^2 + b_k \mathbb{1}}$$

where $a_k, b_k \in \mathbb{R}, k = 1, \ldots, n$ are coefficients that can be optimised to give the best rational approximation of the $sign(H_W)$ with the least number of terms in the right hand side. For example one can use the optimal rational approximation coefficients of Zolotarev [32, 33]. For the rational approximation of $sign(H_W)$ one has:

$$D = \lim_{n \to \infty} S_n$$

In order to compute efficiently the inverse of the preconditioner we go back to the 5-dimensional formulation of the overlap operator (see the previous section as well) which can be written as a matrix in terms of 4-dimensional block matrices:

$$\mathcal{H} = \begin{pmatrix} \frac{1+m_q}{2}\mathbb{1} & \frac{1-m_q}{2}\gamma_5 H_W & \cdots & \frac{1-m_q}{2}\gamma_5 H_W \\ -a_1\mathbb{1} & H_W^2 + b_1\mathbb{1} & & \\ \vdots & & \ddots & \\ -a_n\mathbb{1} & & & H_W^2 + b_n\mathbb{1} \end{pmatrix}.$$

This matrix can be also partitioned in the 2×2 blocked form:

$$\mathcal{H} = \begin{pmatrix} H_{11} & H_{12} \\ H_{21} & H_{22} \end{pmatrix},$$

where the Schur complement is:

$$S_{11} = H_{11} - H_{12} H_{22}^{-1} H_{21}. \qquad (17)$$

We give the following proposition without proof.

Proposition 2. *i) The preconditioner S_n is given by the Schur complement:*

$$S_n = S_{11}$$

ii) Let $\mathcal{H}\chi = \eta$ with $\chi = (y, \chi_1, \ldots, \chi_n)^T$ and $\eta = (r, o, \ldots, o)^T$. Then y is the solution of the linear system $S_n y = r$.

Using these results and keeping n fixed the algorithm of [31] (known also as the multigrid algorithm) has the form of a two level algorithm: This algorithm is in the

---

Algorithm 3   A two-level algorithm for overlap inversion

Set $x^1 \in \mathbb{C}^N$, $r^1 = b$, tol $\in \mathbb{R}_+$, tol$_1 \in \mathbb{R}_+$
for $i = 1, \ldots$ **do**
    Solve approximately $\mathcal{H}\chi^{i+1} = \eta^i$ such that $||\eta^i - \mathcal{H}\chi^{i+1}||_2/||r^i||_2 <$ tol$_1$
    $x^{i+1} = x^i + y^{i+1}$
    $r^{i+1} = b - Dx^{i+1}$
    Stop if $||r^{i+1}||_2/||b||_2 <$ tol
end for

---

form of nested iterations. One can see that the outer loop is Jacobi iteration which contains inside two inner iterations: the approximate solution of the 5-dimensional system and the multiplication with the overlap operator D which involves the computation of the *sign* function. For the 5-dimensional system one can use any iterative solver which suites the properties of $\mathcal{H}$. We have used many forms for $\mathcal{H}$ ranging from rational approximation to domain wall formulations. For the overlap multiplication we have used the algorithm of [17]. Our tests on a small lattice show that the two level algorithm outperforms with an order of magnitude the brute force conjugate gradients nested iterations [31].

Since the inner iteration solves the problem in a 5-dimensional lattice with finite L_5 and the outer iteration solves for the 4-dimensional projected 5-dimensional system with $L_5 \to \infty$, the algorithm in its nature is a multigrid algorithm along the fifth dimension. The fact that the multigrid works here is simply the free propagating fermions in this direction. If this direction is gauged, the usual problems of the multigrid on a 4-dimensional lattice reappear and the idea does not work. In fact, this algorithm with n fixed is a two grid algorithm. However, since it does not involve the classical prolongations and contractions it can be better described as a two-level algorithm. [2]

---

[2] I thank Andreas Frommer for discussions on this algorithm.

References

1. K.G. Wilson, *Confinement of quarks*, Phys. Rev. D 10 (1974) pp. 2445-2459
2. G. W. Wasilkowski and H. Woźniakowski, *On tractability of path integration*, J. Math. Phys. 37 (1994) pp. 2071-2086
3. *Monte Carlo Algorithms for QCD*, this volume.
4. D.H. Weingarten and D.N. Petcher, *Monte Carlo integration for lattice gauge theories with fermions*, Phys. Lett. B99 (1981) 333
5. A. Boriçi and Ph. de Forcrand, *Fast methods for calculation of quark propagator*, Physics Computing 1994, p.711.
6. *Accelerating Wilson fermion matrix inversions by means of the stabilized biconjugate gradient algorithm*, A. Frommer, V. Hannemann, Th. Lippert, B. Noeckel, K. Schilling, Int.J.Mod.Phys. C5 (1994) 1073-1088
7. Z. Bai, M. Fahey, and G. H. Golub, *Some large-scale matrix computation problems*, J. Comp. Appl. Math., 74:71-89, 1996
8. C. Thron, S.J. Dong, K.F. Liu, H.P. Ying, *Padé-Z_2 estimator of determinants* Phys. Rev. D57 (1998) pp. 1642-1653
9. E. Cahill, A. Irving, C. Johnson, J. Sexton, *Numerical stability of Lanczos methods*, Nucl. Phys. Proc. Suppl. 83 (2000) 825-827
10. A. Boriçi, *Computational methods for UV-suppressed fermions*, J. Comp. Phys.189 (2003) 454-462
11. A. Boriçi, *Lattice QCD with suppressed high momentum modes of the Dirac operator*, Phys. Rev. D67(2003) 114501
12. G. Golub, *Variance reduction by control variates in Monte Carlo simulations of large scale matrix functions*, talk given at 3rd Workshop on Numerical Analysis and Lattice QCD, Edinburgh, June 30 - July 4, 2003.
13. A. Boriçi, Global Monte Carlo for light quarks, to be published in the Nuc. Phys. B proceedings of the XXI International Symposium on Lattice Field Theory, Tsukuba, Ibaraki, Japan. See also physics archives `hep-lat/0309044`.
14. A. Boriçi, *Determinant and order statistics*, this volume.
15. A. Boriçi, *On the Neuberger overlap operator*, Phys. Lett. B453 (1999) 46-53
16. A. Boriçi, *Fast methods for computing the Neuberger operator*, in A. Frommer et al (edts.), *Numerical Challenges in Lattice Quantum Chromodynamics*, Springer Verlag, Heidelberg, 2000.
17. A. Boriçi, *A Lanczos approach to the inverse square root of a large and sparse matrix*, J. Comput. Phys. 162 (2000) 123-131
18. C. Lanczos, *Solution of systems of linear equations by minimized iterations*, J. Res. Nat. Bur. Stand., 49 (1952), pp. 33-53
19. P. R. Graves-Morris, *Padé Approximation and its Applications*, Lecture Notes in Mathematics, vol. 765, L. Wuytack, ed.. Springer Verlag, 1979
20. R. W. Freund, *Solution of shifted linear systems by quasi-minimal residual iterations*, in L. Reichel and A. Ruttan and R. S. Varga (edts), Numerical Linear Algebra, W. de Gruyter, 1993
21. G. H. Golub and Z. Strakos, *Estimates in quadratic formulas*, Numerical Algorithms, 8 (1994) pp. 241-268.
22. G. H. Golub and C. F. Van Loan, *Matrix Computations*, The Johns Hopkins University Press, Baltimore, 1989
23. Z. Bai *et al* editors, *Templates for the Solution of Algebraic Eigenvalue Problems: A Practical Guide (Software, Environments, Tools)*, SIAM, 2000.

24. D.B. Kaplan, *A method for simulating chiral fermions on the lattice* Phys. Lett. B 228 (1992) 342.
25. R. Narayanan, H. Neuberger, *Infinitely many regulator fields for chiral fermions*, Phys. Lett. B 302 (1993) 62, *A construction of lattice chiral gauge theories*, Nucl. Phys. B 443 (1995) 305.
26. V. Furman, Y. Shamir, *Axial symmetries in lattice QCD with Kaplan fermions*, Nucl. Phys. B439 (1995) 54-78
27. H. Neuberger, *Exactly massless quarks on the lattice*, Phys. Lett. B 417 (1998) 141
28. A. Boriçi, *Truncated overlap fermions: the link between overlap and domain wall fermions*, in V. Mitrjushkin and G. Schierholz (edts.), Lattice Fermions and Structure of the Vacuum, Kluwer Academic Publishers, 2000.
29. T.-W. Chiu, *Recent developments of domain-wall/overlap fermions for lattice QCD*, Nucl. Phys. B (Proc. Suppl.) Vol. 129-130C (2004) p.135-141
30. A. Boriçi, *Truncated overlap fermions*, Nucl. Phys. Proc. Suppl. 83 (2000) 771-773
31. A. Boriçi, *Chiral fermions and the multigrid algorithm*, Phys. Rev. D62 (2000) 017505
32. E. I. Zolotarev, *Application of the elliptic functions to the problems on the functions of least and most deviation from zero*, Zapiskah Rossijskoi Akad. Nauk. (1877). In Russian.
33. P.P. Petrushev and V.A. Popov, *Rational approximation of real functions*, Cambridge University Press, Cambridge 1987.

Monte Carlo Simulations of Lattice QCD

Mike Peardon

School of Mathematics, Trinity College, Dublin 2, Ireland, mjp@maths.tcd.ie

Summary. This survey reviews computational methodologies in lattice gauge theory as a discretisation of QCD. We particularly focus on techniques for stochastic processes and molecular dynamics which are at the heart of modern lattice QCD simulations.

1 Introduction

Quantum chromodynamics (QCD) is firmly established as the theory describing the strong nuclear force. This force explains how protons and neutrons are bound inside nuclei and how their constituents, the quarks and gluons are permanently confined. QCD is a strongly interacting, asymptotically free quantum field theory and so making predictions for its physical consequences at low energies using the well-established methods of perturbation theory is impossible. The perturbative expansion does not converge in this region. A means of studying the dynamics beyond perturbation theory was devised by Wilson [1] and involves discretising space and time onto a lattice. For introductions to the subject, see Ref. [2] and reviews of the state-of-the-art in the subject can be found in Ref. [3].

The lattice method for computing the properties of quantum field theories begins with the path-integral representation. Our task is to compute observables of the quantum theory of a field $\phi(x)$, given by

$$\langle O \rangle = \frac{1}{\mathcal{Z}} \int \mathcal{D}\phi \; O(\phi) e^{-S(\phi)}.$$

$S(\phi)$ is the action describing the interactions of the theory, $\mathcal{Z}$ is the partition function, chosen such that $\langle 1 \rangle = 1$ and the metric of space-time is assumed to be Euclidean. Most of the properties of the Minkowski space theory can be computed in this metric and as we shall see later, having a theory with a real, positive measure on the space of field configurations, $e^{-S(\phi)} \mathcal{D}\phi$ is crucial if importance sampling Monte Carlo methods are to be employed efficiently.

QCD is a gauge theory, of interacting quarks and gluons. The action involves the Yang-Mills field strength term, and the quark bilinear

$$S_{\text{QCD}} = \int d^4 x \; \frac{1}{2} \text{Tr } F_{\mu\nu} F_{\mu\nu} + \bar{\psi} \left(\gamma_\mu D_\mu + m \right) \psi.$$

A fundamental difficulty arises in writing a lattice representation of the Dirac operator, $\gamma_\mu D_\mu + m$; a naive central difference representation of a first-order derivative causes too many propagating fermion modes to over-populate the lattice. In the Wilson formulation, this problem is cured by adding a higher-dimensional operator (since QCD is asymptotically free, this operator is irrelevant as the lattice spacing is taken to zero). The price for this quick fix is that chiral symmetry is broken. Another form of discretisation leads to the the Kogut-Susskind [4] or staggered lattice fermion. Here, a remnant of chiral symmetry survives and this ensures staggered quarks have many useful properties that make their lattice simulation the cheapest computationally. The difficulty now is that there are still too many flavours, a minimum of four in four dimensions. The effect of fewer flavours is mimicked by taking fractional powers of the staggered determinant, as we will see later. The validity of this trick is still widely debated (see for example Ref. [5]).

The gluon fields are represented on the lattice by link variables, where $U_\mu(x)$ lives on the link in the positive μ direction originating from site x. These links take values in the fundamental representation of the gauge group, $SU(N_c)$ with $N_c = 3$ the number of colours in the theory. For a particular gauge field background, the Wilson fermion matrix is

$$M_W[U]_{xy} = \delta_{xy} - \kappa \sum_\mu U_\mu(x)(1 - \gamma_\mu)\delta_{x+\hat{\mu},y} + U_\mu^\dagger(x - \hat{\mu})(1 + \gamma_\mu)\delta_{x-\hat{\mu},y},$$

where κ is a function of the bare quark mass and γ_μ are the 4×4 Dirac matrices. The staggered fermion matrix is

$$M_S[U]_{xy} = m\delta_{xy} + \frac{1}{2} \sum_\mu \eta_\mu(x)(U_\mu(x)\delta_{x+\hat{\mu},y} - U_\mu^\dagger(x - \hat{\mu})\delta_{x-\hat{\mu},y}).$$

Here, m is the bare quark mass, and $\eta_\mu(x) = (-1)^{x_1 + x_2 + \cdots x_{\mu-1}}$ are the staggered phases. These are independent of the gauge and quark fields and stay fixed in simulations. An important distinction to make at this level is that the two discretisation matrices presented here act on different vector spaces; the staggered fermion has $N_c = 3$ degrees of freedom per lattice site, while the Wilson fermion has $N_c \times 2^{d/2} = 12$ per site.

A more recently developed representation of the fermion operator, the overlap [6] was the centre of attention at this meeting [7]. The lattice fields, like the Wilson action have 12 degrees of freedom per site. This formulation has significant theoretical advantages over the Wilson and staggered formulations, but the advantages come at an algorithmic cost. The new technology needed to simulate overlap quarks is still rapidly developing and so in this introduction, I will focus on the state-of-the-art in simulation techniques that have mostly been developed to simulate simpler formulations of the quark action. These formulations (the Wilson and staggered fermions) couple the quark and gluon fields together through a very sparse matrix. Recall the entries of the fermion matrix depend directly on the gluon fields to be integrated over in the path integral; changing the gluon fields U_μ leads to a change in the matrix $M[U]$. The lattice scheme not only provides a non-perturbative regularisation of the quantum field theory, but in a finite volume, the path integral becomes a finite (but large) dimensional integral. These problems can be tackled numerically by performing Monte Carlo simulations.

2 Importance Sampling Monte Carlo

Importance sampling [8] is a crucial simulation tool in cases when a high-dimensional integral of a function that is sharply peaked in very small regions of the parameter space is being performed. This is certainly the case for the path integral; the action of the field theory is an extensive function of all the degrees of freedom and as such varies by many orders of magnitude as the fields vary. This action is then used to construct a probabilistic Boltzmann weight for each field configuration, $e^{-S(\phi)}\mathcal{D}\phi$ which then varies by "orders-of-magnitudes-of-orders-of-magnitude"! In importance sampling, those regions of the integration volume that contribute most are visited more often in the stochastic sample and so the variance of statistical estimates of the observables of the theory is then greatly reduced.

The use of a stochastic estimate naturally introduces statistical errors into lattice determinations of observables and these must be carefully managed. An ensemble of reasonable size (about 100 configurations as a minimum) must be generated for statistical inference to be reliable. In most modern calculations, these statistical errors are no longer the largest uncertainties, rather it is the cost of generating the ensemble at light fermion masses and on large grid volumes that means extrapolations to the real world must be made. These extrapolations dominate the uncertainties in our calculations: importance sampling is certainly working, but it is extremely expensive.

One other important aspect of Monte Carlo integration of lattice gauge fields is worth mentioning at this stage. The link variables are elements of the gauge group ($SU(3)$ for QCD), which is a compact, curved manifold. The appropriate integration metric on the manifold gives the Haar measure, which has a number of useful properties in group theory that make the lattice formulation of QCD gauge invariant. This integration measure must be taken into account in Monte Carlo simulation, but this seemingly intricate problem is actually solved quite straightforwardly in many of the simulation applications we discuss here.

3 Markov Methods

The problem of generating the required ensemble of configurations is a significant challenge to lattice practitioners. As we shall see later, the action is a complicated function of all the lattice variables and changes to a single variable lead to changes in the action that require computations over the whole set of variables to be performed.

In fact, the only known means of generating an ensemble of configurations suitable for importance sampling calculations is to use a *Markov process*. The Markov process, $\mathcal{M}$ is a memoryless stochastic update which acts on a state of the system to generate a new state. Repeated applications of the process forms a sequence of configurations, the *Markov chain*:

$$\phi_1 \xrightarrow{\mathcal{M}} \phi_2 \xrightarrow{\mathcal{M}} \phi_3 \xrightarrow{\mathcal{M}} \cdots$$

The observation that the Markov process is memoryless means $\mathcal{R}(\phi_{k+1} \longleftarrow \phi_k)$, the probability of jumping from configuration ϕ_k to ϕ_{k+1}, depends only on ϕ_k and ϕ_{k+1} and no other state in the chain; the chain has no recollection of how it came to be in its present state. A few constraints on the Markov process are useful:

firstly the process should be *aperiodic* (that is it does not enter deterministic cycles) and *connected* (there exists a finite number of iterations of the process such that a transition from any state to any other has non-zero probability). A Markov process with these properties is termed *ergodic*.

The important property of any ergodic Markov process is they always tends to some unique, probabilistic fixed point, $\Pi(\phi)$ such that

$$\Pi(\phi) = \int \mathcal{D}\phi' \; \mathcal{R}(\phi \longleftarrow \phi')\Pi(\phi').$$

This fixed-point equation tells us that applying the Markov process once the system has equilibrated generates a new configuration with the same stochastic properties. A particular configuration ϕ has the same chance of appearing before or after the procedure. This suggests a way of building an importance sampling ensemble such that a field occurs with probability $\mathcal{P}(\phi)$; find a Markov process that has $\mathcal{P}$ as its fixed point. Elements in the Markov chain generated by repeated applications of this process then form our ensemble. This method is Markov chain Monte Carlo (MCMC).

The drawback of MCMC is that while the Markov process may be memoryless, the subsequent chain is not. Nearby entries are not necessarily statistically independent. This can be illustrated easily in a toy model. Consider a system that has just two states, $|1\rangle$ and $|2\rangle$ and a Markov process with transitions described by the Markov matrix

$$\mathcal{M} = \begin{pmatrix} 1-\kappa_1 & \kappa_2 \\ \kappa_1 & 1-\kappa_2 \end{pmatrix} \qquad \text{with} \qquad 0 < \kappa_1, \kappa_2 < 1.$$

The matrix entry $\mathcal{M}_{ij}$ is the probability the system will hop to state $|i\rangle$ if it is currently in state $|j\rangle$. The associated process has fixed point probability

$$\Pi = \frac{1}{\kappa_1 + \kappa_2} \begin{pmatrix} \kappa_2 \\ \kappa_1 \end{pmatrix}$$

and so after a large number of iterations, the system is in state $|1\rangle$ with probability $\kappa_2/(\kappa_1+\kappa_2)$ and $|2\rangle$ with probability $\kappa_1/(\kappa_1+\kappa_2)$. The conditional probability that the system is in state $|i\rangle$ after t iterations, given it was in state $|j\rangle$ initially can be computed as $[\mathcal{M}^N]_{ij}$ so for example when $i = j = 1$, the probability is

$$P(\phi(t) = |1\rangle | \phi(0) = |1\rangle) = \frac{\kappa_2}{\kappa_1 + \kappa_2} + \frac{\kappa_1(1 - \kappa_1 - \kappa_2)^t}{\kappa_1 + \kappa_2}. \tag{1}$$

Since $|1 - \kappa_1 - \kappa_2| < 1$, the second term in Eqn. 1 decreases by a factor of $\lambda_2 = 1 - \kappa_1 - \kappa_2$ every iteration and λ_2 is the second largest eigenvalue of the Markov matrix. The first term is just the probability the system is in state $|1\rangle$ in the fixed-point distribution and the exponential term is the memory of the Markov chain. These statistical dependencies between nearby entries are called *autocorrelations*. Update algorithms are not then all born equal; two different algorithms can have the same fixed point, but very different autocorrelations. The better algorithm will have a Markov matrix with smaller second and higher eigenvalues. A perfect "heat-bath" method would have all other eigenvalues equal to zero.

No perfect algorithms have been constructed for QCD, so all Markov chains used in simulations will have autocorrelations. These must be monitored carefully in the analysis stage of the lattice calculation. Measurements of the autocorrelations can be made by looking at the statistics of elements of the chain at increasing separations and testing for independence. In simple averages, data can be gathered into increasingly large bins until statistical error estimates converge.

The most widely used method to construct a Markov process with a particular fixed point is to use *detailed balance*. If a Markov process obeys detailed balance for fixed point $\mathcal{P}$, the transition rates $\mathcal{R}_\mathcal{P}$ obey

$$\mathcal{R}_\mathcal{P}(\phi' \longleftarrow \phi)\mathcal{P}(\phi) = \mathcal{R}_\mathcal{P}(\phi \longleftarrow \phi')\mathcal{P}(\phi'),$$

and since

$$\int \mathcal{D}\phi' \mathcal{R}_\mathcal{P}(\phi' \longleftarrow \phi) = 1,$$

$\mathcal{R}_\mathcal{P}$ has fixed point $\mathcal{P}$ as required. Detailed balance is a stronger constraint than is necessary to ensure the Markov process has the desired fixed point, however it is a very useful tool in practice. If two or more algorithms that have a common fixed point are combined to form a compound process, the resulting method shares the same fixed point, although it does not necessarily obey detailed balance for that fixed point. The best known examples of Markov processes constructed to obey detailed balance are those based on the *Metropolis algorithm* [9].

Metropolis algorithms have two components: a *proposal* step and an *accept/reject* test. In the proposal step, a reversible and area-preserving map on the configuration space is applied to the current element at the end of the chain. For this map,

$$\mathcal{R}_1(\phi' \longleftarrow \phi) = \mathcal{R}_1(\phi \longleftarrow \phi').$$

At the second stage, ϕ' is randomly assigned as the new configuration in the chain with probability, $\mathcal{P}_\text{acc}$ given by the Metropolis test:

$$\mathcal{P}_\text{acc} = \min\left[1, \frac{\mathcal{P}(\phi')}{\mathcal{P}(\phi)}\right].$$

If the proposal ϕ' is rejected, the current entry ϕ is duplicated in the chain. The Metropolis test has the useful property that the overall normalisation of the probabilities of the two states, ϕ and ϕ' is not required, only their ratio. For the exponential weight used in lattice field theory Monte Carlo simulations, this ratio is a function of the difference in the action on the two configurations: $\mathcal{P}(\phi')/\mathcal{P}(\phi) = \exp\left(S(\phi) - S(\phi')\right)$.

4 Including the Dynamics of Quark Fields

Quarks are fermions. To represent a fermion field in a path integral, the Grassmann algebra is used. Grassmann variables anti-commute and have simple integration rules,

$$\chi\eta = -\eta\chi, \qquad \int d\chi = 0 \quad \text{and} \quad \int \chi\, d\chi = 1. \tag{2}$$

Naturally, these types of variables can not be stored directly in a computer, and while in principle all the required properties of the algebra can be evaluated numerically, this requires prohibitively large amounts of storage for practical calculation. Happily, the quark fields of QCD interact with the gluons via a quark bilinear operator and the fermion path integral can be computed analytically on a given gluon background. The integral of a function of many Grassmann variables can be computed using the rules of Eqn. 2, taking care to keep track of the sign flips from the anticommutation property. The path integral result, for N_f flavours of mass-degenerate quarks is

$$\int \prod_f \mathcal{D}\bar{\psi}_f \mathcal{D}\psi_f \; \exp \sum_f^{N_f} \bar{\psi}_f M[U] \psi_f = (\det M[U])^{N_f}.$$

A full derivation is presented in most quantum field theory texts (See *e.g.* Chapter 9 of Ref. [10]). For QCD, asymptotic freedom tells us that operators containing four or more quark fields are irrelevant in the continuum limit. Quartic terms can however appear in improved operators, designed to reduce the effects of the finite lattice cut-off. The difficulties they introduce into simulations mean lattice physicists have tended to avoid their use. They can be handled in simulation by introducing extra auxiliary fields that must be integrated over as part of the Monte Carlo calculation. In nature, the two lightest flavours of quark, "up" and "down" are almost massless (compared to the scales relevant to QCD interactions) so most effort in algorithm development has focused on simulating a pair of quark flavours. This situation is in fact simpler than studying QCD with a single flavour of quark, as we shall see. The determinants are often expressed as an effective action, so that $\det M[U]^{N_f} = e^{-S_{\text{eff}}[U]}$ with

$$S_{\text{eff}}[U] = -N_f \ln \det M[U]. \tag{3}$$

The QCD lattice path integral is then expressed as

$$Z = \int \mathcal{D}U \mathcal{D}\bar{\psi} \mathcal{D}\psi \; e^{-S_G[U] + \bar{\psi} M[U] \psi} = \int \mathcal{D}U \; e^{-S_G[U] - S_{\text{eff}}[U]}.$$

The drawback with solving the fermion path integral directly is that now we have a non-local function on the gluon field. Here, non-local means that computing the change in the effective action of Eqn. 3 when a single link variable is altered requires at least $\mathcal{O}(V)$ operations (where V is the number of degrees of freedom in the simulation). This is in contrast to simulations of the $SU(3)$ Yang-Mills theory, where this change in the action requires $\mathcal{O}(1)$ computations and very efficient simulation algorithms that change a single degree of freedom at a time can be constructed.

The most widely used method in importance sampling of including the fermion determinant when there is an even number of quark fields (or flavours) is to reintroduce an integral representation of the determinant, but this time in terms of standard complex variables. These variables are usually called the *pseudofermions* [11]. For two flavours of quarks, the path integral is

$$\det M[U]^2 = \det M^\dagger M = \int \mathcal{D}\phi^* \mathcal{D}\phi \; e^{-\phi^* [M^\dagger M]^{-1} \phi}, \tag{4}$$

where a useful property of the fermion matrix M has been used, namely

$$M^\dagger[U] = \gamma_5 M[U] \gamma_5.$$

This is usually termed "γ_5-hermiticity" and implies $\det M^\dagger = \det M$ (since $\det \gamma_5 = 1$). The result of Eqn. 4 is easily derived by noting that if the positive-definite matrix $M^\dagger M$ is diagonalised, the result is a product of a large set of simple Gaussian integrals, with widths related to the eigenvalues of $M^\dagger M$. Note that this description is still non-local.

An alternative representation of the fermion path integral was proposed by Lüscher [12]. Starting with $P(x)$, an order n polynomial approximation to $1/x$,

$$\frac{1}{x} \approx P(x) = \sum_{n=0}^{n} c_n x^n = c_n \prod_{i=1}^{n} (x - z_i),$$

where $\{z_1, z_2, \ldots z_n\}$ are the n roots of the polynomial. The polynomial of the positive-definite matrix, $Q^2 = X^\dagger X$ is then an approximation to the inverse of Q^2, so

$$Q^{-2} \approx P(Q^2) \propto \prod_{i=1}^{n} (Q - \mu_i)(Q - \mu_i^*).$$

The weights μ_i are related to the roots of the polynomial, $\mu_i = \sqrt{z_i}$. The two flavour fermion determinant can now be represented as a Gaussian integral of a large set of newly introduced fields, χ_k with a local action

$$\det Q^2 \approx \frac{1}{\det P(Q^2)} \propto \int \prod_{i=1}^{n} \mathcal{D}\chi_i \mathcal{D}\chi_i^* \ \exp\left(-\sum_k \chi_k^*(Q-\mu_k)(Q-\mu_k^*)\chi_k\right).$$

The locality of the action means the local update methods that proved to be so effective for simulations of the Yang-Mills theory of gluons alone can be reused. Using a finite polynomial means the algorithm is approximate, becoming exact only when the degree of the polynomial becomes infinite. The problem discovered with the method is that, while QCD is only recovered as $n \to \infty$, the local updates that can be performed become "stiff" [13]. The change in a link variable after an update becomes smaller and smaller as n increases. Also, as the fermion mass is decreased toward its physical value, the order of the polynomial required for a certain level of precision must increase.

A number of proposals to make this formulation exact for any degree of polynomial have been made [14, 15, 16]. These methods work by introducing an accept-reject step to make this correction. In his original paper, Lüscher suggested a re-weighting of the configurations in the ensemble to correct for the approximation. These techniques have been combined and have become quite sophisticated [17]. They are used in large-scale production runs and can be adapted to cope with an odd number of fermion flavours.

5 Molecular Dynamics

The most widely used algorithms in large-scale production runs for QCD calculations are based on the ideas of molecular dynamics. These ideas were first developed from stochastic differential equations such as the Langevin equation [18, 19] and then hybrid forms appeared [20]. The advantage of molecular dynamics is that it is deterministic and so suffers less from random-walk autocorrelations. For a comparison and discussion, see Ref. [21]. As a first step, a trivial modification to the probability density is made by simply adding a set of Gaussian stochastic variables and forming the joint probability density,

$$\mathcal{P}_\mathcal{J}(\pi,\phi) = \frac{1}{Z_J} e^{-\frac{1}{2}\pi^2 - S(\phi)}.$$

Since this joint distribution is separable, expectation values of functions on the original fields of the theory, ϕ are completely unchanged by the modification. A fictitious, "simulation time" coordinate, τ is then introduced and a Hamiltonian is defined,

$$\mathcal{H} = \frac{1}{2}\pi^2 + S(\phi). \tag{5}$$

This formulation now suggests that the newly introduced variables, π should be considered as conjugate momenta to ϕ for evolution in τ. The equations of motion for this dynamics are easily evaluated as

$$\frac{d\phi}{d\tau} = \pi,$$
$$\frac{d\pi}{d\tau} = -\frac{\partial S}{d\phi}. \tag{6}$$

Integrating these equations of motion defines a mapping, $\{\phi,\pi\} \longrightarrow \{\phi',\pi'\}$ which exactly conserves the Hamiltonian of Eqn. 5 and Liouville's theorem implies it is area preserving. The mapping is also reversible (once we introduce a momentum flip), since if $\{\phi,\pi\} \longrightarrow \{\phi',\pi'\}$ then $\{\phi',-\pi'\} \longrightarrow \{\phi,-\pi\}$. This tells us the mapping formed by integrating the equations of motion of Eqn. 6 forms a Markov process with the correct fixed point. This process is not however ergodic. The Hamiltonian $\mathcal{H}$ is conserved, so only contours of phase space with the same value of $\mathcal{H}$ as the original state of the system can be visited. A simple remedy can be made; since the joint probability distribution of the system is separable, the momenta, π can be refreshed occasionally from a heat-bath. This ensures all the contours corresponding to all values of $\mathcal{H}$ can be explored.

For simulations of QCD the fields on the lattice are elements of the fundamental representation of the gauge group, $\mathcal{G} = SU(3)$. In the molecular dynamics approach, the lattice fields are considered to be coordinates of particles moving in an abstract space. If these fields are then elements of a compact group, the molecular dynamics must describe motion of points on a curved manifold, rather than a flat, non-compact space. The properties of Lie groups tell us how to define this motion, since small changes to a group element are described by elements of the Lie algebra. With $g \in \mathcal{G}, \pi \in \mathcal{L}(\mathcal{G})$, a small change can be written

$$g(\tau) \longrightarrow g(\tau+h) = e^{ih\pi}g(t) \approx (1+ih\pi)g(\tau).$$

and so
$$\frac{dg}{d\tau} = i\pi g.$$

These observations provide a natural way to define a canonical momentum conjugate to a coordinate of a group element. For the fundamental representation of $SU(N)$, the Lie algebra is the vector space of traceless, hermitian $N \times N$ matrices. The molecular dynamics coordinates are the link variables, $U_\mu(x) \in \mathcal{G}$ and the momenta are then $p_\mu(x) \in \mathcal{L}(\mathcal{G})$. Time evolution for the gauge fields is then

$$U(\tau + h) = e^{ih\pi(\tau)} U(\tau).$$

A similar "right momentum" could be defined using the convention $\dot{U} = iU\pi$. The interacting Hamiltonian of Eqn. 5 is non-linear and analytic solutions certainly seem impossible to find. As a result, a numerical integration of the equations of motion defined by the Hamiltonian (Eqn. 6) is required, using a finite-step-size integrator.

A number of algorithms based around molecular dynamics (and similarly Langevin evolution) have been developed. One of the most widely used is the *R-algorithm* which is able to handle fractional powers of the fermion matrix. This property is useful for simulations of an arbitrary number of flavours of staggered quarks (although the interpretation of the fractional power of the determinant as a rigorous quantum field theory is still debated). In the R-algorithm, rather than representing the fermionic part of the QCD action with a Gaussian integral, a stochastic estimator of the matrix inverse is constructed and used to generate the force on the momenta. The construction of the algorithm ensures that, by evaluating the matrix at appropriately chosen points in the numerical integration, the finite-step-size errors appear at $\mathcal{O}(h^2)$. The use of numerical integration still means the algorithm is inexact and control of the $h \to 0$ limit is required.

6 Exact Algorithms

Molecular dynamics integrators with a finite step-size do not exactly conserve the Hamiltonian. One commonly used integration method, the leap-frog scheme is however exactly area-preserving and reversible (at least in exact arithmetic). The leap-frog integration scheme with step-size h is

$$\phi(t) \longrightarrow \phi(t + h/2) = \phi(t) + \frac{h}{2}\pi(t),$$

$$\pi(t) \longrightarrow \pi(t + h) = \pi(t) - h\frac{\partial S}{\partial \phi}(t + h/2),$$

$$\phi(t + h/2) \longrightarrow \phi(t + h) = \phi(t + h/2) + \frac{h}{2}\pi(t + h).$$

The reversibility of the leapfrog scheme is seen from the construction; note that the force term, $\frac{\partial S}{\partial \phi}$ is computed using the field ϕ evaluated at the half-step, $t + h/2$. Each of the three components of the scheme is area-preserving since only one of the two conjugate degrees of freedom is changed during a given step. Since the three components have unit Jacobian, so does the composite.

In section 3 it was noted that an area-preserving, reversible proposal scheme is precisely the required first component of the Metropolis algorithm. This observation results in an exact algorithm for simulation with dynamical fermions, the Hybrid Monte Carlo (HMC) algorithm [22]. A leap-frog integrator is used to provide a proposal to a Metropolis test, and ergodicity is maintained by refreshing the momenta (and any fields representing the fermions) between the Metropolis test and the next molecular dynamics stage.

The original formulation of the algorithm used the pseudofermion description of the quark fields. The Hamiltonian for evolution in the fictitious time co-ordinate, τ is

$$\mathcal{H} = \frac{1}{2}\pi^2 + S_G[U] + \phi^* \left[M(U)^\dagger M(U)\right]^{-1} \phi.$$

To ensure ergodicity, a compound Markov process is again used. At the beginning of each update, the pseudofermion fields and the momenta are updated from a Gaussian heat-bath. The pseudofermion fields ϕ are held fixed during the molecular dynamics evolution.

For the pseudofermion action, $S_{\mathrm{pf}} = \phi^* \left[M(U)^\dagger M(U)\right]^{-1} \phi$, the subsequent force term on the momenta is evaluated by considering

$$\frac{dS}{d\tau} = -\phi^* \left[M^\dagger M\right]^{-1} \left(\frac{dM^\dagger}{d\tau}M + M^\dagger \frac{dM}{d\tau}\right) \left[M^\dagger M\right]^{-1} \phi$$

$$= -Y^* \frac{dM}{d\tau} X - X^* \frac{dM^\dagger}{d\tau} Y$$

with

$$Y = M^{\dagger -1}\phi \quad \text{and} \quad X = M^{-1}Y. \tag{7}$$

The most significant expense for numerical simulation arises from the evaluation of this pseudofermion force term. This requires the vectors X and Y of Eqn. 7 to be solved at each step, as the gauge fields change (and hence M changes). The physical parameter region of real interest to lattice physicists is the *chiral limit*, $m_q \to 0$. In this limit the fermion matrix becomes increasingly ill-conditioned and inversion becomes more costly. If the inversion process can be accelerated, then the Monte Carlo algorithm is similarly accelerated. As a result of this observation, a good deal of effort in the community has gone into finding the best inversion method for the fermion operator. For the more established fermion methods, the matrix is very sparse, so finding optimisations for iterative Krylov-space methods was the focus of this effort. For staggered quarks, the conjugate gradient algorithm applied to $M^\dagger M$ works very well (this is equivalent to using an even-odd preconditioning). For the Wilson fermion matrix, more intensive studies [23] have found BiCGStab is optimal. With operators such as the overlap, which is a dense matrix, finding fast inversion algorithms is a new challenge, discussed at length at the meeting.

Krylov-space methods can be improved greatly by an efficient preconditioner. For both staggered and Wilson fermions, the preconditioning most widely discussed is the even-odd (or red-black) scheme. This is an Incomplete Lower-Upper (ILU) factorisation, constructed by first indexing the sites of the lattice so even sites ($x + y + z + t \bmod 2 = 0$) are ordered first. Now, the fermion matrix can be written

$$M = \begin{pmatrix} M_{ee} & M_{eo} \\ M_{oe} & M_{oo} \end{pmatrix}$$

and for both Wilson and staggered quarks, M_{ee} and M_{oo} are easily inverted, so the ILU factorisation (equivalent to a Schur decomposition) gives

$$M = \begin{pmatrix} I_{ee} & 0 \\ M_{ee}^{-1}M_{oe} & I_{oo} \end{pmatrix} \begin{pmatrix} M_{ee} & 0 \\ 0 & M_{oo} - M_{oe}M_{ee}^{-1}M_{eo} \end{pmatrix} \begin{pmatrix} I_{ee} & M_{oe}^{-1}M_{eo} \\ 0 & I_{oo} \end{pmatrix}.$$

The inversion of M proceeds by finding the inverse of the sub-lattice operator $M_{oo} - M_{oe}M_{ee}^{-1}M_{eo}$ on a modified source. ILU factorisation has been extended to full lattice lexicographic ordering of the sites. Here, better acceleration of the fermion inversion algorithms is seen, but these methods are harder to implement on parallel computers. A compromise was developed in Ref. [24], with the development of locally lexicographic schemes (sites stored on a single processor are ordered).

Preconditioning such as the red-black scheme can be incorporated into the matrix appearing in the pseudofermion action: any positive matrix A for which $\det A[U] = \det M^\dagger[U]M[U]$ is suitable, and ILU preconditioning preserves this property. The red-black method [25] is commonly used, and other preconditioners have been explored [26, 27] with some speed-ups seen.

After the local bosonic representation of the fermion determinant was suggested, Frezzotti and Jansen proposed using this formulation within HMC, giving the PHMC algorithm [28]. They recognised a useful property of polynomials, namely that they provided a useful means of making modest changes to the importance sampling measure that include modes with smaller eigenvalues. These modes often make significant contributions to expectation values of interesting physical processes.

7 Recent Developments

The ultimate goal of lattice simulation is to approach the point where quarks are (almost) massless. This is termed the chiral limit; chirality is an exact symmetry of the quark fields in the limit $m_q \to 0$. The physical up and down quark masses are very close to this limit, certainly lighter than lattice simulations have been able to reach to date.

The variance in the stochastic estimator of the fermion action, for example the pseudofermion fields, diverges as the chiral limit is approached. A recent suggestion by Hasenbusch [29] to reduce the variance in the estimate of the quark dynamics is to split the pseudofermion fields into two pieces, and use the Gaussian integral representation separately for each new piece. The identity

$$\det M^2 = \det(M\tilde{M}^{-1})^2 \det \tilde{M}^2 = \int \mathcal{D}\phi\mathcal{D}\phi^*\mathcal{D}\chi\mathcal{D}\chi^* e^{-S_{\text{split}}},$$

with

$$S_{\text{split}} = \phi^* \tilde{M} \left[M^\dagger M \right]^{-1} \tilde{M}^\dagger \phi + \chi^* \left[\tilde{M}^\dagger \tilde{M} \right]^{-1} \chi,$$

then follows, for any non-singular matrix, $\tilde{M}$. The advantage of the method is the new stochastic variables introduced to represent the fermions can be coupled to the gauge fields through two better-conditioned matrices, $\tilde{M}$ and $M\tilde{M}^{-1}$. The molecular dynamics step-size can be increased by a significant factor for a fixed Metropolis acceptance and this factor is seen to increase in the chiral limit, $m_q \to 0$. The idea can be generalised and a longer chain of S splittings can be made, since

$$\det M = \det M\tilde{M}_1^{-1} \det \tilde{M}_1\tilde{M}_2^{-1} \det \tilde{M}_2\tilde{M}_3^{-1} \cdots \det \tilde{M}_S.$$

At present an interesting question remains; is the "infinite split" limit of the algorithm, into $S \to \infty$ better-conditioned fragments, equivalent to using the effective action of Eqn. 3 directly, without introducing the extra noise associated with a stochastic estimate?

Another means of accelerating the molecular-dynamics evolution was suggested by Sexton and Weingarten [30]. They noted that generalised leap-frog integrators could be constructed with different effective integration step-sizes for different segments of the action. The method was tested by splitting the action into the gauge and pseudofermionic parts, but proved to be of limited use. The method is only useful if two criteria can be met simultaneously. Firstly, the splitting must break the action into a piece whose force is cheap to compute and a harder-to-evaluate piece and secondly, the cheap force segment must have the most rapidly fluctuating molecular dynamics modes. A low-order polynomial in the PHMC algorithm has been proposed [31] as a simple way of performing this splitting, and preliminary tests are encouraging.

The real world of QCD is recovered in the continuum limit, as the lattice spacing is taken to zero. Renormalisation arguments tell us that there are infinitely many lattice theories all of which describe QCD in this limit. This idea is known as *universality* and means that not only can different fermion formalisms be used, but also that the gluon and quark field interactions can be introduced in different ways. The important constraint in designing a lattice description of QCD is to maintain gauge invariance of the theory; without this, universality arguments are severely weakened and the lattice theory does not necessarily describe the right theory. There have been a number of developments leading to new means of coupling the quark and gluon fields in ways that reproduce the continuum limit more rapidly. The key idea is to reduce the coupling of the very short distance modes of the theory, which should play a minor role in the low-energy dynamics we are interested in. These reduction methods are termed *smearing* [32]. The recent use of these methods in lattice simulations stemmed from observations that dislocations (abrupt changes to the gauge fields) could be greatly reduced by applying a smearing algorithm to the gauge fields. The link variables appearing in the quark matrix, M are directly replaced by the smeared links. Discretisations built from these links are called "fat-link" actions [33]. Fat-link simulations have been performed using global Metropolis methods [34]: remarkably, these algorithms work well once the short-distance modes are tamed. Fat-link methods suitable for use in molecular-dynamics schemes are now being introduced [35]. Recently [36], a fully differentiable smearing method, the "stout-link" was proposed that is better suited to molecular-dynamics based algorithms.

Horváth, Kennedy and Sint [37] proposed an exact algorithm for directly simulating any number of quark flavours. The algorithm begins with a rational approximation to a fractional power α, e.g

$$x^{-\alpha} \approx R(x) = \frac{A(x)}{B(x)},$$

where $A(x)$ and $B(x)$ are polynomials of degree r. Clearly, the inverse of this approximation is just $B(x)/A(x)$. A pseudofermion representation of any power of the determinant, $\det M^\alpha$ is then constructed by first noting

$$|\det M|^{2\alpha} = \det(M^\dagger M)^\alpha = \det R^\dagger(M) R(M)$$

and then using a Gaussian integral representation in the standard way. Multi-shift solvers [38] allow for efficient application of the matrices A/B or B/A, once these polynomial ratios are written as a partial fraction sum. The development of this algorithm, along with PHMC, seems to provide attractive solutions to the use of inexact algorithms, such as the popular R-algorithm for simulations with a single flavour of quark or two staggered flavours. These algorithms are currently being tested [39] in large-scale production runs.

8 Conclusions

The numerical study of quantum field theories with lattice simulations has evolved significantly over the past twenty years, but a number of its foundation stones remain unchanged. All lattice simulations are based on Markov chain Monte Carlo and all studies involving fermions are significantly more computationally expensive than those with bosonic fields alone. The expense of simulations with quarks is the major obstacle and one that never ceases to surprise theorists since the fermion statistics of Eqn. 2 are much simpler than boson statistics.

The algorithmic problem considered by this meeting is to develop the toolkit of algorithms further to encompass the newly-devised fermion formulations that obey the Ginsparg-Wilson relation.

References

1. K. G. Wilson, Phys. Rev. D **10** (1974) 2445.
2. A number of good introductory texts are available. See "Lattice gauge theories: an introduction", H. .J. Rothe (World Scientific), "Quantum fields on a lattice", I. Montvay and G. Munster (Cambridge) and "Introduction to Quantum fields on a lattice", J. Smit (Cambridge) for example.
3. A. Frommer and H. Neuberger, these proceedings. Also, see for example T. De-Grand [arXiv:hep-ph/0312241] for a recent comprehensive review.
4. J. B. Kogut and L. Susskind, Phys. Rev. D **11** (1975) 395. L. Susskind, Phys. Rev. D **16** (1977) 3031.
5. K. Jansen, arXiv:hep-lat/0311039.
6. H. Neuberger, Phys. Lett. B **417** (1998) 141.
7. H. Neuberger, arXiv:hep-lat/0311040.
8. "Monte Carlo Methods", Hammersley and Handscomb (Chapman and Hall).
9. N. Metropolis, A. Rosenbluth, M. Rosenbluth, A. Teller and E. Teller, J. Chem. Phys. **21** (1953) 1087.
10. "An Introduction to Quantum Field Theory" M. Peskin and D. Schroeder (Addison Wesley).
11. D. H. Weingarten and D. N. Petcher, Phys. Lett. B **99** (1981) 333.
12. M. Luscher, Nucl. Phys. B **418** (1994) 637.
13. B. Bunk, K. Jansen, B. Jegerlehner, M. Luscher, H. Simma and R. Sommer, Nucl. Phys. Proc. Suppl. **42** (1995) 49.
14. M.. Peardon [UKQCD Collaboration], Nucl. Phys. Proc. Suppl. **42** (1995) 891
15. A. Borici and P. de Forcrand, Nucl. Phys. B **454** (1995) 645
16. I. Montvay, Nucl. Phys. B **466** (1996) 259
17. C. Gebert and I. Montvay, arXiv:hep-lat/0302025.
18. A. Ukawa and M. Fukugita, Phys. Rev. Lett. **55** (1985) 1854.
19. G. G. Batrouni, G. R. Katz, A. S. Kronfeld, G. P. Lepage, B. Svetitsky and K. G. Wilson, Phys. Rev. D **32** (1985) 2736.
20. S. Duane, Nucl. Phys. B **257** (1985) 652.
21. D. Toussaint, Comput. Phys. Commun. **56** (1989) 69.
22. S. Duane, A. D. Kennedy, B. J. Pendleton and D. Roweth, Phys. Lett. B **195** (1987) 216.

23. A. Frommer, V. Hannemann, B. Nockel, T. Lippert and K. Schilling, Int. J. Mod. Phys. C **5** (1994) 1073
24. S. Fischer, A. Frommer, U. Glassner, T. Lippert, G. Ritzenhofer and K. Schilling, Comput. Phys. Commun. **98** (1996) 20
25. R. Gupta, A. Patel, C. F. Baillie, G. Guralnik, G. W. Kilcup and S. R. Sharpe, Phys. Rev. **D40** (1989) 2072.
26. P. de Forcrand and T. Takaishi, Nucl. Phys. Proc. Suppl. **53** (1997) 968.
27. M. J. Peardon, arXiv:hep-lat/0011080.
28. R. Frezzotti and K. Jansen, Phys. Lett. B **402** (1997) 328.
29. M. Hasenbusch, Phys. Lett. B **519** (2001) 177.
30. J. C. Sexton and D. H. Weingarten, Nucl. Phys. B **380** (1992) 665.
31. M. J. Peardon and J. Sexton, Nucl. Phys. Proc. Suppl. **119**, 985 (2003).
32. M. Albanese *et al.* [APE Collaboration], Phys. Lett. B **192** (1987) 163.
33. T. DeGrand [MILC Collaboration], Phys. Rev. D **58** (1998) 094503.
34. A. Hasenfratz and F. Knechtli, Comput. Phys. Commun. **148** (2002) 81.
35. W. Kamleh, D. B. Leinweber and A. G. Williams, arXiv:hep-lat/0309154.
36. C. Morningstar and M. J. Peardon, arXiv:hep-lat/0311018.
37. A. D. Kennedy, I. Horvath and S. Sint, Nucl. Phys. Proc. Suppl. **73** (1999) 834.
38. A. Frommer, B. Nockel, S. Gusken, T. Lippert and K. Schilling, Int. J. Mod. Phys. C **6** (1995) 627.
39. M. A. Clark and A. D. Kennedy, arXiv:hep-lat/0309084.

Part II

Lattice QCD

Determinant and Order Statistics

Artan Boriçi

School of Physics, The University of Edinburgh, James Clerk Maxwell Building, Mayfield Road, Edinburgh EH9 3JZ, borici@ph.ed.ac.uk

Summary. Noisy methods give in general a biased estimation of the quark determinant. In this paper we describe order statistics estimators which eliminate this bias. The method is illustrated in case of Schwinger model on the lattice.

1 Introduction

The most difficult task in light quark simulations is the computation of the quark determinant. It is well-known that the determinant of a matrix $A \in \mathbb{C}^{N \times N}$ scales poorly with N and may also be ill-conditioned for ill-conditioned matrices A [1]. This is the case for lattice QCD with light quarks. Hence, Monte Carlo algorithms are bound to produce gauge fields which are closely related to each other. Therefore, to get the next independent gauge field configuration we must run a large number of Monte Carlo steps, each one requiring the computation of the determinant or its variation. This way efficient methods are sought for the quark determinant computation.

The standard method is the Gaussian integral representation. If A is positive definite and Hermitian one has:

$$\det A = \int \prod_{i=1}^{N} \frac{d\mathrm{Re}(\phi_i) d\mathrm{Im}(\phi_i)}{\pi} \; e^{-\phi^H A^{-1} \phi}$$

If A represents the change between two underlying gauge fields one is often interested in its noisy estimation:

$$\det A \sim e^{\phi^H (\mathbb{1} - A^{-1}) \phi}$$

One can use iterative algorithms to invert A which are well suited for large and sparse problems and have $O(N)$ complexity.

However, such an estimator may not adequately represent determinant changes. To see this consider $A = \mathrm{diag}(\varepsilon, 1/\varepsilon)$ with $\varepsilon \in (0, 1)$. The stochastic estimator of the determinant gives:

$$\det A \sim e^{(1-\varepsilon^{-1})|\phi_1|^2+(1-\varepsilon)|\phi_2|^2}$$

Note that the right hand side goes to zero for $\varepsilon \to 0$ while $\det A = 1$ for any $\varepsilon \in (0,1)$.

Noisy methods improve this situation. They are based on the identity:

$$\det A = e^{Tr \log A}$$

and the noisy estimation of the trace of the natural logarithm of A [2, 8, 4, 5].

Let $Z_j \in \{+1, -1\}, j = 1, \ldots, N$ be independent and identically distributed random variables with probabilities:

$$\operatorname{Prob}(Z_j = 1) = \operatorname{Prob}(Z_j = -1) = \frac{1}{2}, \quad j = 1, \ldots, N$$

Then for the expectation values we get:

$$\mathbb{E}(Z_j) = 0, \quad \mathbb{E}(Z_j Z_k) = \delta_{jk}, \quad j, k = 1, \ldots, N$$

and the following result holds:

Proposition 1. Let X be a random variable defined by:

$$X = -Z^T \log A Z, \quad Z^T = (Z_1, Z_2, \ldots, Z_N)$$

Then its expectation μ and variance σ^2 are given by:

$$\mu = \mathbb{E}(X) = -Tr \log A, \quad \sigma^2 = \mathbb{E}[(X-\mu)^2] = 2\sum_{j \neq k}[Re(\log A)_{jk}]^2$$

Proof. To simplify notations we define $B = -\log A$ and $Y = X - \mu$. then we have:

$$X = TrB + \sum_{i \neq j} Z_i Z_j B_{ij}$$

It is easy to show that:

$$Z_{ij} = Z_{ji} = Z_i Z_j, \quad Z_{ij} \in \{+1, -1\}$$

where:

$$\operatorname{Prob}(Z_{ij} = 1) = \operatorname{Prob}(Z_{ij} = -1) = \frac{1}{2}$$

Then we can write:

$$Y = \sum_{i<j} Z_{ij}(B_{ij} + B_{ji}) \equiv \sum_{i<j} Y_{ij}$$

where

$$\mathbb{E}(Y_{ij}) = 0, \quad \mathbb{E}(Y_{ij}^2) = (B_{ij} + B_{ji})^2$$

Since B is Hermitian then $B_{ij} + B_{ji} = 2ReB_{ij}$, so we get:

$$\mathbb{E}(Y) = 0, \quad \mathbb{E}(Y^2) = 2\sum_{i \neq j}(ReB_{ji})^2 \quad \square$$

To evaluate the matrix logarithm one can use the methods described in [2, 3, 4, 5]. These methods have similar complexity with the inversion algorithms and have been reviewed by the same author in this volume [6].

Now let us turn to the 2×2 example where $A = \mathrm{diag}(\varepsilon, 1/\varepsilon)$ with $\varepsilon \in (0,1)$. This means that we are looking for a noisy estimator of the trace of the diagonal matrix $-\log A = \mathrm{diag}(-\log \varepsilon, \log \varepsilon)$. In this case the noisy estimator of the determinant is given by:
$$\det A \sim e^{-\log \varepsilon (Z_1^2 - Z_2^2)} = e^0 = 1$$
Hence, the noisy estimator is not spoiled by the ill-conditioning of the matrix A.

However, this method gives in general a biased estimation of the determinant as it is clear from Jensen's inequality $\mathbb{E}(e^{-X}) \geq e^{\mathbb{E}(-X)}$. To compute the bias we use the central limit theorem:

Proposition 2. *For N large the random variable X is normally distributed with mean μ and variance σ^2.*

Proof. The proof is based on the fact that $Y = X - \mu$ is a sum of $N(N-1)/2$ independent and identically distributed variables with mean zero and variance $(2 Re B_{ij})^2$. □

Using this result the expected value of the noisy exponential is given by:
$$\mathbb{E}(e^{-X}) = e^{-\mu + \sigma^2/2}$$

Clearly, the bias can be reduced by reducing the variance of the estimation. If the estimator is the average $\bar{X}$ of the sample $X_1, \ldots, X_n$ then we get:
$$\mathbb{E}(e^{-\bar{X}}) = e^{-\mu + \sigma^2/(2n)}$$

This method is not practical since the sample volume grows proportionally to σ^2 which itself grows proportionally to N (see Proposition (1)).

There are many variance reduction techniques: [3] subtract traceless matrices which results in an error reduction from 559% to 17%; [7] proposes a control variate technique which can be found in this volume; [8] suppresses large eigenvalues of the fermion determinant, a formulation that reduces by an order of magnitude unphysical variations in the determinant. All these improvements reduce significantly the variance. Nonetheless, its linear scaling with N doesn't change.

In this paper we discuss an asymptotically normal estimator which gives an unbiased determinant. Such an estimator is shown to be a central order statistic.

2 Distribution of an Order Statistic

It is well-known that the sample median is an estimator of the population mean. If $X_1, X_2, \ldots, X_n$ is a sample drawn from the same distribution, the median is defined as the middle point of the ordered sample:
$$X_{(1)} \leq X_{(2)} \leq \ldots \leq X_{(n)}$$

This is possible if the sample size n is odd. In case n is even the median is defined to be any number in the interval $[X_{(n/2)}, X_{(n/2+1)}]$.

Median is a particular example of a statistic calculated from the ordered data. In general, $X_{(k)}, k = 1, \ldots, n$ is called the k'th *ordered statistic* [9], [10].

Let $X_1, X_2, \ldots, X_n$ be distributed according to $f(x), x \in R$. Distribution of an order statistic $X_{(k)}$ is denoted by $f_k(x)$. In this case $f(x)$ is called the *parent distribution*.

To find $f_k(x)$ one can use a simple heuristic. First, let $f_k(x)dx$ be the probability that $X_{(k)}$ is in an infinitesimal interval dx about x. This means that one of the sample variables is in this interval, $k-1$ are less than x and $n-k$ are greater than x. The number of ways of choosing these variables is the multinomial coefficient $C(n; k-1, 1, n-k)$. If $F(x)$ is the cumulative distribution function of the sample variables then one gets:

$$f_k(x)dx = C(n; k-1, 1, n-k)[F(x)]^{k-1}[1 - F(x)]^{n-k} f(x)dx$$

which gives:

$$f_k(x) = \frac{n!}{(k-1)!(n-k)!}[F(x)]^{k-1}[1 - F(x)]^{n-k} f(x) \qquad (1)$$

The same result can be obtained more formally. Let $F_k(x)$ be the cumulative distribution function of $X_{(k)}$:

$$F_k(x) = \text{Prob}\{X_{(k)} \leq x\}$$

This is the probability that at least k of X_k are less or equal to x which is given by the cumulative distribution function of the binomial distribution:

$$F_k(x) = \sum_{j=k}^{n} C_n^j [F(x)]^j [1 - F(x)]^{n-j}$$

Differentiating with respect to x we have:

$$f_k(x) = \sum_{j=k}^{n} [T_{j-1}(x) - T_j(x)], \quad T_j(x) = (n-j) C_n^j [F(x)]^j [1 - F(x)]^{n-j-1} f(x)$$

Since $T_n(x) = 0$ the sum telescopes down to $T_{k-1}(x)$, which is the same as equation (1).

3 Asymptotic Theory

The distribution of an order statistic can be of little use for direct analytical calculations of moments for most of known parent distributions. Of course one can use numerical methods to compute numerical values to the desired accuracy. However, as the sample size grows the order statistic distribution approaches some limiting distribution. The situation is analogous to the central limit theorem. In general the distribution is not necessarily asymptotically normal as one could have expected although this will be the case for the *central order statistics* that will be dealt with below.

For $0 < \alpha < 1$, let $i_\alpha = [n\alpha] + 1$, where $[n\alpha]$ represents the integer part of $n\alpha$. Let also F be absolutely continuous with f positive at $F^{-1}(\alpha)$ and continuous at that point. Then the following result is true:

Theorem 1. *(asymptotic distribution of a central order statistic).* As $n \to \infty$,

$$\sqrt{n} f[F^{-1}(\alpha)] \frac{X_{(i_\alpha)} - F^{-1}(\alpha)}{\sqrt{\alpha(1-\alpha)}} \xrightarrow{d} N(0,1)$$

For the proof see [10].

Example. Suppose that the distribution is symmetric around the population mean μ. Let also assume that the variance σ^2 is finite and $f(\mu)$ is finite and positive. For simplicity n is taken to be odd. Then the median $X_{(n+1)/2}$ is an unbiased estimator of μ and asymptotically normal. Further, $\text{Var}[X_{(n+1)/2}] \approx [4nf(\mu)^2]^{-1}$.

4 Unbiased Estimation of the Fermion Determinant

Let the $X_i, i = 1, ..., n$ be a normally distributed sample of unbiased estimators of $-Tr \log A$. We are interested in the expected value of $e^{-X_{(i_\alpha)}}$ with $i_\alpha = [n\alpha] + 1$ and α as above. Now let $x_\alpha := (F^{-1}(\alpha) - \mu)/\sigma$. Then $X_{(i_\alpha)}$ is asymptotically normal with mean $\mu + x_\alpha \sigma$ and variance $2\pi\sigma^2 \alpha(1-\alpha)e^{x_\alpha^2}/n$. For the expected value one gets:

$$\mathbb{E}[e^{-X_{(i_\alpha)}}] = e^{-\mu - x_\alpha \sigma + \pi\sigma^2 \alpha(1-\alpha)e^{x_\alpha^2}/n}$$

To have an unbiased estimator of the fermion determinant in terms of an ordered statistic one should have:

$$\mathbb{E}[e^{-X_{(i_\alpha)}}] = e^{-\mu}$$

This can be achieved if:

$$x_\alpha = \pi\sigma\alpha(1-\alpha)e^{x_\alpha^2}/n$$

The crucial question here is: given n and σ is there a solution of the above equality in terms of x_α? To answer this question we express first α in terms of x_α. Since $(X - \mu)/\sigma \sim N(0,1)$ then we have:

$$\alpha = P[(X-\mu)/\sigma < x_\alpha] = \frac{1}{2} + \frac{1}{2}\text{erf}(\frac{x_\alpha}{\sqrt{2}})$$

Hence, the equation to be solved is:

$$x_\alpha = \frac{\pi\sigma}{4n}[1 - \text{erf}(\frac{x_\alpha}{\sqrt{2}})^2]e^{x_\alpha^2} \qquad (2)$$

Using Taylor expansion around $x_\alpha = 0$ one gets:

$$x_\alpha = \frac{\pi\sigma}{4n}[1 + O(x_\alpha^2)]$$

For small x_α the $O(x_\alpha^2)$ can be neglected. One gets a unique solution:

$$x_\alpha = \frac{\pi\sigma}{4n}, \qquad x_\alpha \ll 1 \qquad (3)$$

In this case the sample size should be much greater then σ and i can be calculated to give:

$$i = [n\alpha] + 1 = [\frac{n}{2} + \sqrt{\frac{\pi}{2}\frac{\sigma}{4}}] + 1$$

Thus, using order statistics it is possible to have an unbiased estimator of the fermion determinant from a biased high estimator of the fermion effective action.

However, the estimator is only asymptotically unbiased meaning that the sample size has to be large enough. In practice, at least from the examples below the asymptotic regime is reached for $n = 40$. In many application to lattice fermions σ is large anyway. Thus, requiring $n \gg \sigma$ will satisfy the conditions of the asymptotic normality as well.

Note that equation (2) may have more than one solution. Here we restrict ourselves to the small x_α solution (3). This solution must be made accurate using it as a starting guess in a non-linear equation solver.

Since σ is unknown one has to estimate it from the sample. This is a source of systematic error that may bias the stochastic estimation of the determinant. To compute the error one has to substitute σ with its estimator S in equation (2). This way we get a solution x_β and the error given by:

$$\mathbb{E}[e^{\mu - X_{(i_\beta)}}] = e^{-x_\beta \sigma + \frac{\pi \sigma^2}{4n}[1-\mathrm{erf}(\frac{x_\beta}{\sqrt{2}})^2]e^{x_\beta^2}} = e^{x_\beta \sigma(\frac{\sigma}{S}-1)} \quad (4)$$

In order to control this bias one has to compute the right hand side and make sure that it is small. This can be done using a large sample so that σ is accurately determined. If possible, one should cross check the results with other methods as it is shown in the next section.

5 Schwinger Model on the Lattice

Schwinger model is the theory of quantum electrodynamics in two dimensions (QED2). When discretized on a lattice it shares many features of lattice QCD. Therefore it is a good starting point to test computational methods, which otherwise require huge computing resources when applied directly to QCD.

Let Λ be the collection of points $x = (x_1, x_2)$ on a square lattice in two dimensions. Let also $U(x)_\mu, x \in \Lambda$ be an unimodular complex number, an element of the $U(1)$ group. It is a map onto $U(1)$ gauge group of the oriented link connecting lattice sites x and $x + ae_\mu$. If $\phi(x), x \in \Lambda$ is a complex valued function on the lattice, then covariant differences are defined by:

$$\nabla_\mu \phi(x) = \frac{1}{2a}[U(x)_\mu \phi(x + ae_\mu) - U^H(x - ae_\mu)_\mu \phi(x - ae_\mu)]$$

where a and e_μ are the lattice spacing and the unit lattice vector along the coordinate $\mu = 1, 2$.

There are two main discretizations of the Dirac operator on the lattice: the Wilson and Kogut-Susskind operators. Here we use the latter since it is easier to manipulate. It is the matrix operator defined by:

$$[D_{KS}(m, U)\phi](x) = m\phi(x) - \sum_{\mu=1}^{2} \eta(x)_\mu \nabla_\mu \phi(x)$$

where m is the electron mass and

$$\eta(x)_1 = 1, \quad \eta(x)_2 = (-1)^{x_1}$$

The total number of lattice points is $N = L_1 L_2$ on a lattice with L_1, L_2 sites in each dimension.

Note that $D_{KS}(m,U)$ is a non-Hermitian operator. Let Σ be a diagonal matrix with $\Sigma_{x,x} = (-1)^{x_1+x_2}$, i.e. its elements change sign according to a checkerboard or even-odd partition of lattice sites. Since $D_{KS}(0,U)$ links even and odd sites, it anticommutes with Σ:

$$D_{KS}(0,U)\Sigma + \Sigma D_{KS}(0,U) = 0$$

This way it is possible to define a Hermitian operator:

$$H_{KS}(m,U) = \Sigma D_{KS}(m,U)$$

The gauge action which describes the dynamics of the gauge field is given by:

$$S_g(U) = \beta Re \sum_{\mathbb{P}} (\mathbb{1} - U_{\mathbb{P}})$$

where $\mathbb{P}$ denotes the elementary square on the lattice or the plaquette. The sum in the right hand side is over all plaquettes. $U_{\mathbb{P}}$ is a $U(1)$ element defined on the plaquette $\mathbb{P}$ and $\beta = 1/(e^2 a^2)$ is the coupling constant of the theory, e being the electron charge.

To solve the Schwinger model on the lattice one has to evaluate the path integral:

$$Z = \int \prod_{x \in \Lambda} dU(x)_1 dU(x)_2 \, [\det D_W(m,U)]^{N_f} e^{-S_g(U)}$$

where N_f is the number of electrons in the system. We take here $N_f = 2$. Using the fact that $\det \Sigma = 1$ one has:

$$\det D_{KS}(m,U) = \det H_{KS}(m,U)$$

To simulate the model we use the Metropolis *et al* algorithm [11] in two steps. First we propose local changes in the gauge field using the gauge action alone such that the detailed balance is satisfied:

$$e^{-[S_g(U') - S_g(U)]} P_g(U \to U') = P_g(U' \to U)$$

By selecting randomly the next lattice gauge field, one Monte Carlo sweep consists of applying the above algorithm at each lattice site once. To ensure that gauge field proposals are statistically independent we perform 10000 Monte Carlo sweeps and check autocorrelations of the time series of the measured data. Then the second step is to accept the new configuration U' with the Metropolis *et al* probability:

$$\hat{P}_{acc}[X_{(i_\alpha)}; U, U'] = \min\{1, e^{-X_{(i_\beta)}}\}$$

with

$$\frac{\det H_{KS}(m,U')^2}{\det H_{KS}(m,U)^2} \equiv e^{\mu(U,U')} = \mathbb{E}[e^{-X_{(i_\alpha)}}] \approx \mathbb{E}[e^{-X_{(i_\beta)}}] \quad (5)$$

where $X_{(i_\alpha)}$ and $X_{(i_\beta)}$ are the order statistics from the sample $X_1, \ldots, X_n$ of unbiased noisy estimators of:

$$\mu(U,U') = -Tr \log[A(m,U') - A(m,U)] \quad (6)$$

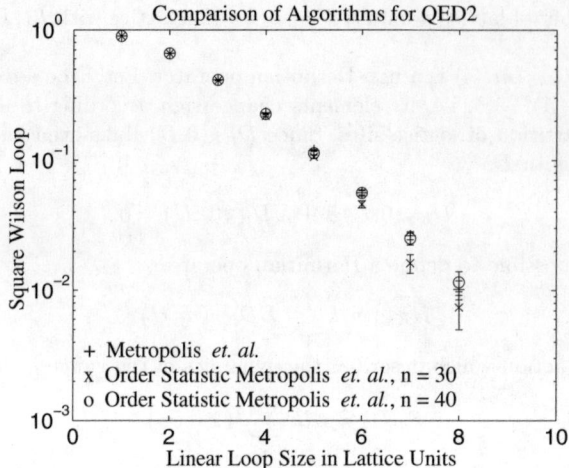

Fig. 1. Two flavors of Swchinger model on a 16 × 16 lattice at $\beta = 5$ and bare quark Kogut-Susskind fermion mass $m = 0.01$

and α, β are defined in the previous section.

The average acceptance probability is defined by:

$$P_{acc}(U, U') = \int_{-\infty}^{\infty} dx f_{i_\alpha}(x; U, U') \min\{1, e^{-x}\} \quad (7)$$

Using this definition it can be shown that the detailed balance is satisfied:

$$e^{\mu(U,U')} P_{acc}(U, U') = P_{acc}(U', U) \quad (8)$$

The proof is given in the appendix.

Having an unbiased determinant as in (5), the above algorithm is not a surprise. What is worth noting here is that this algorithm works for gauge field changes of any size unlike the generally accepted view on such algorithms [12].

To illustrate the algorithm we measure Wilson loops on a 16 × 16 lattice at $\beta = 5$ and bare fermion mass $m = 0.01$. In Figure 1 we show Wilson loops as a function of the linear size of the loop using three algorithms: 1) computing exactly the determinant ratios, 2) using order statistics with $n = 30$ noisy estimators and 3) the same as in 2) but with $n = 40$ noisy estimators.

Note that the exact computation of the determinant is a $O(N^3)$ process, which is very demanding for four dimensional models. This complexity should be compared to the $O(Nn)$ complexity of the noisy method. Using eq. (4) of the previous section we find that for a fixed error one must have $n \sim N^{2/3}$. This result can be derived by noting that $x_\beta \sim \sigma/n$, $\sigma^2 \sim N$ and that $\mathbb{E}(\sigma - S) \sim n^{-1/2}$. Hence, the complexity of the method is $O(N^{1.67})$. The acceptance of the Metropolis *et al* for the exact ratio was about 16%, whereas the acceptance of the new algorithm was about 20%.

We see that already $n = 40$ estimators are enough to reach the exact ratio result. This example illustrates the validity of the algorithm and that one can use different sample volumes in order to control systematic errors. These errors come from deviations from normality of the order statistic distribution and inaccurate estimations of variance σ^2. Since σ^2 grows proportionally to lattice volume N, one can argue that deviations from normality should be small on larger lattices. The more difficult error to control is the error on variance estimation. To check the error one can repeat calculations for different sample volumes like in the example above. But this may be expensive and a cheaper way is to calculate the error directly using eq. (4) of the previous section.

We plan to test the algorithm in the future in the case of QCD with UV-suppressed fermions [8].

References

1. *Bounds on the trace of the inverse and the determinant of symmetric positive definite matrices*, Annals of Numer. Math. 4 (1997) pp. 29-38
2. Z. Bai, M. Fahey, and G. H. Golub, *Some large-scale matrix computation problems*, J. Comp. Appl. Math., 74:71-89, 1996
3. C. Thron, S.J. Dong, K.F. Liu, H.P. Ying, *Padé-Z_2 estimator of determinants* Phys. Rev. D57 (1998) pp. 1642-1653
4. E. Cahill, A. Irving, C. Johnson, J. Sexton, *Numerical stability of Lanczos methods*, Nucl. Phys. Proc. Suppl. 83 (2000) 825-827
5. A. Boriçi, *Computational Methods for UV-Suppressed Fermions*, J. Comp. Phys.189 (2003) 454-462
6. A. Boriçi, *Computational methods for the fermion determinant and the link between overlap and domain wall fermions*, this volume.
7. G. Golub, *Variance Reduction by Control Variates in Monte Carlo Simulations of Large Scale Matrix Functions*, talk given at 3rd Workshop on Numerical Analysis and Lattice QCD, Edinburgh, June 30 - July 4, 2003.
8. A. Boriçi, *Lattice QCD with Suppressed High Momentum Modes of the Dirac Operator*, Phys. Rev. D67(2003) 114501
9. H.A. David, *Order Statistics*, John Wiley & Sons, Inc., 1970
10. B. C. Arnold, N. Balakrishnan, H. N. Nagaraja, *A First Course in Order Statistics*, John Wiley & Sons, Inc., 1992
11. *Equation of state calculations by fast computing machines*, N. Metropolis, A.W. Rosenbluth, M.N. Rosenbluth, A.H. Teller, E. Teller, J.Chem.Phys.21 (1953) 1087-1092
12. A.D. Kennedy, J. Kuti *Noise without noise: a new Monte Carlo method*, Phys. Rev. Lett. 54 (1985) 2473

A Proof for Detailed Balance

From the normality of $X \sim N[\mu(U, U'), \sigma(U, U')]$ we have:

$$F(x; U, U') = \frac{1}{2} + \frac{1}{2}\mathrm{erf}\{[x - \mu(U, U')]/[\sqrt{2}\sigma(U, U')]\}$$

and

$$f(x; U, U') = \frac{1}{\sqrt{2\pi}\sigma(U, U')} e^{[x-\mu(U,U')]^2/[2\sigma(U,U')^2]}$$

Substituting these in the order statistics distribution (1) and setting $k = i_\alpha$ the acceptance probability (7) is given by:

$$P_{acc}(U, U') =$$

$$\int_{-\infty}^{\infty} dx \, kC_n^k \, [F(x; U, U')]^{k-1}[1 - F(x; U, U')]^{n-k} f(x; U, U') \min\{1, e^{-x}\}$$

Changing the integration variable to $y = F(x; U, U')$ we get:

$$P_{acc}(U, U') = \int_0^1 dy \, kC_n^k \, y^{k-1}(1-y)^{n-k} \min\{1, e^{-[\mu(U,U') + \sqrt{2}\sigma \operatorname{erf}^{-1}(2y-1)]}\}$$

Then the reverse acceptance probability will be:

$$P_{acc}(U', U) = \int_0^1 dy \, kC_n^k \, y^{k-1}(1-y)^{n-k} \min\{1, e^{-[\mu(U',U) + \sqrt{2}\sigma \operatorname{erf}^{-1}(2y-1)]}\}$$

Using $\mu(U', U) = -\mu(U, U')$ (see (6)) and changing the integration variable once more to $y \to (1-y)$ we get:

$$P_{acc}(U', U) = \int_0^1 dy \, kC_n^k \, (1-y)^{k-1} y^{n-k} \min\{1, e^{\mu(U,U') + \sqrt{2}\sigma \operatorname{erf}^{-1}(2y-1)}\} \quad (9)$$

Now it is easy to check detailed balance (8). There are two cases:
i) For $\mu(U, U') + \sqrt{2}\sigma \operatorname{erf}^{-1}(2y-1) \leq 0$ the left hand side of (8) gives $e^{\mu(U,U')}$ whereas using (9) the right hand side of (8) is:

$$P_{acc}(U', U) = e^{\mu(U,U')} \int_0^1 dy \, kC_n^k \, (1-y)^{k-1} y^{n-k} e^{\sqrt{2}\sigma \operatorname{erf}^{-1}(2y-1)}$$

Then changing the integration variable again, $y \to (1-y)$, we get:

$$P_{acc}(U', U) =$$

$$e^{\mu(U,U')} \int_0^1 dy \, kC_n^k \, y^{k-1}(1-y)^{n-k} e^{-\sqrt{2}\sigma \operatorname{erf}^{-1}(2y-1)} = e^{\mu(U,U')} \mathbb{E}[e^{-X_{(k)}}]$$

where $X \sim N(0, \sigma)$. But the expectation of the order statistic $\mathbb{E}[e^{-X_{(k)}}] = 1$ in this case and $k = i_\alpha$.
ii) For $\mu(U, U') + \sqrt{2}\sigma \operatorname{erf}^{-1}(2y-1) > 0$ the left hand side of (8) is:

$$e^{\mu(U,U')} P_{acc}(U, U') =$$

$$\int_0^1 dy \, kC_n^k \, y^{k-1}(1-y)^{n-k} e^{-\sqrt{2}\sigma \operatorname{erf}^{-1}(2y-1)} = \mathbb{E}[e^{-X_{(k)}}]$$

Again, $\mathbb{E}[e^{-X_{(k)}}] = 1$ for $X \sim N(0, \sigma)$ and $k = i_\alpha$. Using (9), the right hand side of (8) gives $\mathbb{E}[1] = 1$. $\square$

Monte Carlo Overrelaxation for $SU(N)$ Gauge Theories

Philippe de Forcrand[1,2] and Oliver Jahn[1,3]

[1] Institute for Theoretical Physics, ETH, CH-8093 Zürich, Switzerland
[2] CERN, Physics Department, TH Unit, CH-1211 Geneva 23, Switzerland
forcrand@phys.ethz.ch
[3] Center for Theoretical Physics, MIT, Cambridge, MA 02139, USA jahn@mit.edu

Summary. The standard approach to Monte Carlo simulations of $SU(N)$ Yang-Mills theories updates successive $SU(2)$ subgroups of each $SU(N)$ link. We follow up on an old proposal of Creutz, to perform overrelaxation in the full $SU(N)$ group instead, and show that it is more efficient.

The main bottleneck in Monte Carlo simulations of QCD is the inclusion of light dynamical fermions. For this reason, algorithms for the simulation of Yang-Mills theories have received less attention. The usual combination of Cabibbo-Marinari pseudo-heatbath [1] and Brown-Woch microcanonical overrelaxation [2] of $SU(2)$ subgroups is considered satisfactory. However, the large-N limit of QCD presents a different perspective. Fermionic contributions are suppressed as $1/N$, so that studying the large-N limit of Yang-Mills theories is interesting in itself. High precision is necessary to isolate not only the $N \to \infty$ limit, but also the leading $1/N$ correction. Such quantitative studies by several groups are underway [3]. They show that dimensionless combinations of the glueball masses, the deconfinement temperature T_c, and the string tension σ approach their $N \to \infty$ limit rapidly, with rather small corrections $\sim 1/N^2$, even down to $N = 2$. The prospect of making $\mathcal{O}(1/N^2) \sim 10\%$, or even $\mathcal{O}(1/N) \sim 30\%$ accurate predictions for real-world QCD is tantalizing. Numerical simulations can guide theory and help determine the $N = \infty$ "master field". Already, an old string prediction $T_c/\sqrt{\sigma}(N=\infty) = \sqrt{\frac{3}{\pi(d-2)}}$, first dismissed by the author himself because it disagreed with Monte Carlo data at the time [4], appears to be accurate to within 1% or better. Proposals about the force between charges of k units of Z_N charge, so-called k-string tensions, can be confronted with numerical simulations, which may or may not give support to connections between QCD and supersymmetric theories [5]. Efficient algorithms for $SU(N)$ Yang-Mills theories are highly desirable.

Here, we revive an old, abandoned proposal of Creutz [6], to perform overrelaxation in the full $SU(N)$ group, and show its superiority over the traditional $SU(2)$ subgroup approach[4].

---
[4] Global updates, of Hybrid Monte Carlo type, are not considered here, because they were shown to be $\mathcal{O}(100)$ times less efficient than local ones for $SU(3)$ in [7].

1 State of the Art

We consider the problem of updating a link matrix $U \in SU(N)$, from an old value U_old to U_new, according to the probability density

$$P(U) \propto \exp(\beta \frac{1}{N} \operatorname{Re} \operatorname{Tr} X^\dagger U) \quad . \tag{1}$$

$\frac{1}{N} \operatorname{Re} \operatorname{Tr} X^\dagger U$ is the "local action". The matrix X represents the sum of the "staples", the neighboring links which form with U a closed loop contributing to the action. This is the situation for the Wilson plaquette action, or for improved actions (Symanzik, Iwasaki, ...) containing a sum of loops all in the fundamental representation. Higher representations make the local action non-linear in U. This typically restricts the choice of algorithm to Metropolis, although the approach below can still be used to construct a Metropolis candidate (as e.g. in [8]). Thus, X is a sum of $SU(N)$ matrices, i.e. a general $N \times N$ complex matrix.

Three types of local Monte Carlo algorithms have been proposed:
- **Metropolis**: a random step R in $SU(N)$ is proposed, then accepted or rejected. Thus, from U_old, a candidate $U_\text{new} = RU_\text{old}$ is constructed. To preserve detailed balance, the Metropolis acceptance probability is

$$P_\text{acc} = \min(1, \exp(\beta \frac{1}{N} \operatorname{Re} \operatorname{Tr} X^\dagger (U_\text{new} - U_\text{old}))) \quad . \tag{2}$$

Acceptance decreases as the stepsize, measured by the deviation of R from the identity, increases. And an $N \times N$ matrix multiplication must be performed to construct U_new, which requires $\mathcal{O}(N^3)$ operations. This algorithm is simple but inefficient, because the direction of the stepsize is random. By carefully choosing this direction, a much larger step can be taken as we will see.
- **Heatbath**: a new matrix U_new is generated directly from the probability density $P(U)$ Eq.(1). This is a manifest improvement over Metropolis, since U_new is completely independent of U_old. However, sampling $P(U)$ requires knowledge of the normalization on the right-hand side of Eq.(1). For $SU(2)$, the simple algorithm of [9] has been perfected for large β [10]. For $SU(3)$, a heatbath algorithm also exists [11], although it can hardly be called practical. For $SU(N), N > 2$, one performs instead a pseudo-heatbath [1]. Namely, the matrix U_old is multiplied by an embedded $SU(2)$ matrix $R = \mathbf{1}_{N-2} \otimes R_{SU(2)}$, chosen by $SU(2)$ heatbath from the resulting probability $\propto \exp(\beta \frac{1}{N} \operatorname{Re} \operatorname{Tr}(X^\dagger U_\text{old})R)$. Note that computation of the 4 relevant matrix elements of $(X^\dagger U_\text{old})$ requires $\mathcal{O}(N)$ work. To approach a real heatbath and decrease the correlation of $U_\text{new} = U_\text{old} R$ with U_old, a sequence of $SU(2)$ pseudo-heatbaths is usually performed, where the $SU(2)$ subgroup sweeps the $\frac{N(N-1)}{2}$ natural choices of off-diagonal elements of $\tilde{U}$. The resulting amount of work is then $\mathcal{O}(N^3)$, which remains constant relative to the computation of X as N increases.
- **Overrelaxation**. Adler introduced stochastic overrelaxation for multi-quadratic actions [12]. The idea is to go beyond the minimum of the local action and multiply this step by $\omega \in [1, 2]$, "reflecting" the link U_old with respect to the action minimum. This results in faster decorrelation, just like it produces faster convergence in linear systems. In fact, as in the latter, infrared modes are accelerated at the expense of ultraviolet modes, as explained in [13]. The overrelaxation parameter ω can be tuned. Its optimal value approaches 2 as the dynamics becomes more critical. In this limit, the UV modes do not evolve and the local action is conserved, making

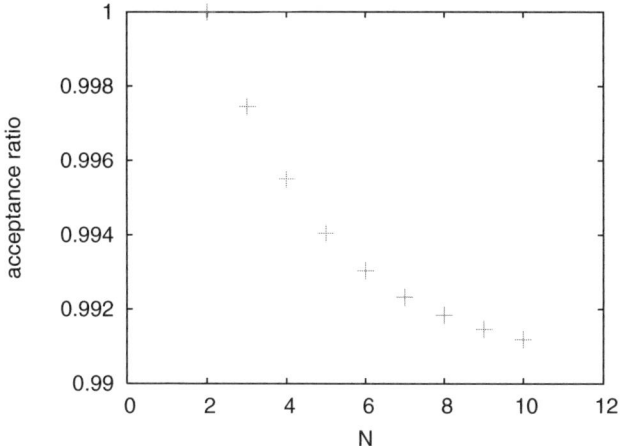

Fig. 1. $SU(N)$ overrelaxation acceptance versus N.

the algorithm microcanonical. In practice, it is simpler to fix ω to 2, and alternate overrelaxation with pseudo-heatbath in a tunable proportion (typically 1 HB for 4-10 OR, the latter number increasing with the correlation length). In $SU(2)$, this strategy has been shown to decorrelate large Wilson loops much faster than a heatbath [14], with in addition some slight reduction in the amount of work. It is now the adopted standard. For $SU(N)$, one performs $\omega = 2$ microcanonical overrelaxation steps in most or all $SU(2)$ subgroups, as described in [2].

The $SU(2)$ subgroup overrelaxation of Brown and Woch is simple and elegant. Moreover, it requires minimal changes to an existing pseudo-heatbath program. But it is not the only possibility. Creutz [6] proposed a general overrelaxation in the $SU(N)$ group. And Patel [15] implemented overrelaxation in $SU(3)$, whose efficiency was demonstrated in [7]. Here, we generalize Patel's method to $SU(N)$.

2 $SU(N)$ Overrelaxation

It may seem surprising at first that working with $SU(N)$ matrices can be as efficient as working on $SU(2)$ subgroups. One must bear in mind that the calculation of the "staple" matrix X requires $\mathcal{O}(N^3)$ operations, since it involves multiplying $N \times N$ matrices. The relative cost of updating U will remain bounded as N increases, if it does not exceed $\mathcal{O}(N^3)$ operations. An update of lesser complexity will use a negligible fraction of time for large N, and can be viewed as a wasteful use of the staple matrix X. Therefore, it is a reasonable strategy to spend $\mathcal{O}(N^3)$ operations on the link update. A comparison of efficiency should then be performed between (i) an update of all $\frac{N(N-1)}{2}$ $SU(2)$ subgroups, one after the other, following Cabibbo-Marinari and Brown-Woch; (ii) a full $SU(N)$ update, described below, involving a polar decomposition of similar $\mathcal{O}(N^3)$ complexity. One may still worry that (ii) is unwise because the final acceptance of the proposed U_{new} will decrease very fast as N increases. Fig. 1 addresses this concern: the acceptance of our $SU(N)$ update scheme decreases in fact very slowly with N, and remains almost 1 for all practical N values.

We now explain how to perform $SU(N)$ overrelaxation, along the lines of [6]. The idea of overrelaxation is to go, in group space, in the direction which minimizes the action, but to go beyond the minimum, to the mirror image of the starting point. If $\hat{X}$ is the $SU(N)$ group element which minimizes the action, then the rotation from U_{old} to $\hat{X}$ is $(\hat{X} U_{\text{old}}^{-1})$. Overrelaxation consists of applying this rotation twice:

$$U_{\text{new}} = (\hat{X} U_{\text{old}}^{-1})^2 U_{\text{old}} = \hat{X} U_{\text{old}}^{\dagger} \hat{X} \qquad (3)$$

U_{new} should then be accepted with the Metropolis probability Eq.(2). The transformation Eq.(3) from U_{old} to U_{new} is an involution (it is equal to its inverse). From this property, detailed balance follows.

Note that this holds for *any* choice of $\hat{X}$ which is independent of $U_{\text{new/old}}$, resulting always in a valid update algorithm. Its efficiency, however, depends on making a clever choice for $\hat{X}$. The simplest one is $\hat{X} = \mathbf{1}\ \forall \mathbf{X}$, but the acceptance is small. Better alternatives, which we have tried, build $\hat{X}$ from the Gram-Schmidt orthogonalization of X, or from its polar decomposition. We have also considered applying Gram-Schmidt or polar decomposition to $X^{\dagger}$ or to X^*. In all cases, a subtle issue is to make sure that U_{new} is indeed special unitary ($\det U_{\text{new}} = 1$), which entails cancelling in $\hat{X}$ the phase usually present in $\det X$. The best choice for $\hat{X}$ balances work, Metropolis acceptance and effective stepsize. Our numerical experiments have led us to the algorithm below, based on the polar decomposition of X, which comes very close to finding the $SU(N)$ matrix which minimizes the local action. Note that Narayanan and Neuberger [16] have converged independently to almost the same method (they do not take Step 3 below).

2.1 Algorithm

1. Perform the Singular Value Decomposition (SVD) of X: $X = U \Sigma V^{\dagger}$, where U and $V \in U(N)$, and Σ is the diagonal matrix of singular values σ_i ($\sigma_i = \sqrt{\lambda_i}$, where the λ_i's are the eigenvalues of the non-negative Hermitian matrix $X^{\dagger} X$). It is simple to show that $W \equiv UV^{\dagger}$ is the $U(N)$ matrix which maximizes $\operatorname{Re}\operatorname{Tr} X^{\dagger} W$. Unfortunately, $\det UV^{\dagger} \neq 1$.
2. Compute $\det X \equiv \rho \exp(i\phi)$. The matrix $\hat{X}_{NN} = \exp(-i\frac{\phi}{N}) UV^{\dagger}$ is a suitable $SU(N)$ matrix, adopted by Narayanan and Neuberger [16].
3. Find an approximate solution $\{\theta_i\}$ for the phases of the diagonal matrix $D = \operatorname{diag}(\exp(i\theta_1), .., \exp(i\theta_N)), \sum_N \theta_i = 0 \bmod 2\pi$, to maximize the quantity $\operatorname{Re}\operatorname{Tr} X^{\dagger}(\exp(-i\frac{\phi}{N}) UDV^{\dagger})$. To find an approximate solution to this non-linear problem, we assume that all phases θ_i are small, and solve the linearized problem.
4. Accept the candidate $U_{\text{new}} = \hat{X} U_{\text{old}}^{\dagger} \hat{X}$, where $\hat{X} = \exp(-i\frac{\phi}{N}) UDV^{\dagger}$, with probability Eq.(2) [5].

2.2 Efficiency

We set out to compare the efficiency of the algorithm above with that of the standard $SU(2)$ subgroup approach, as a function of N. Going up to $N = 10$ forced us to

---
[5]This corresponds to an overrelaxation parameter $\omega = 2$. It may be possible to make the algorithm more efficient by tuning ω, using the LHMC approach of [17].

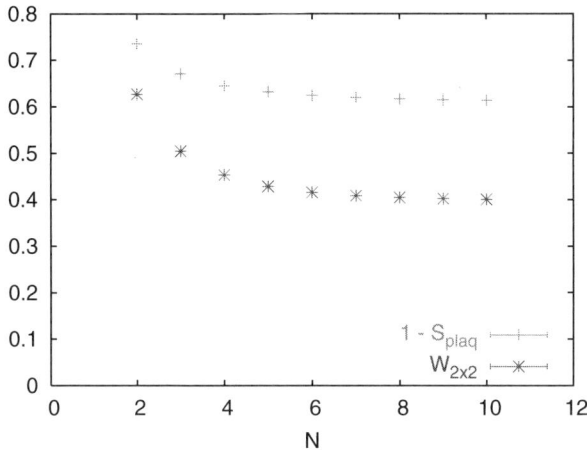

Fig. 2. Expectation values of 1x1 and 2x2 Wilson loops, versus N, for $\beta = \frac{3}{4}N^2$.

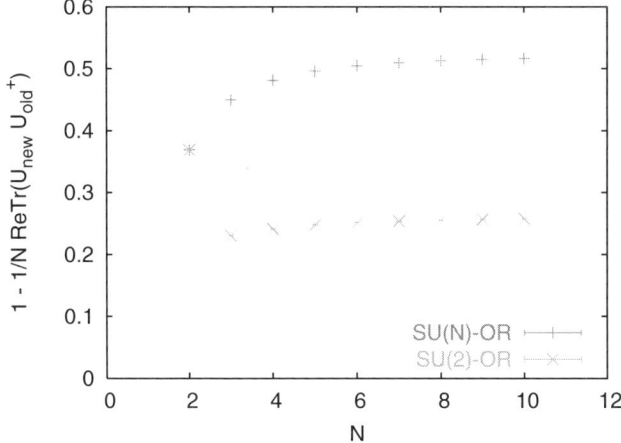

Fig. 3. Distance in group space after one link update, $\langle 1 - \frac{1}{N} \operatorname{Re} \operatorname{Tr} U_{\text{new}}^\dagger U_{\text{old}} \rangle$ vs N.

consider a very small, 2^4 system only. We chose a fixed 't Hooft coupling $g^2 N = 8/3$, so that $\beta = \frac{2N}{g^2} = \frac{3}{4}N^2$. This choice gives Wilson loop values varying smoothly with N, as shown in Fig. 2, and is representative of current $SU(N)$ studies. The Metropolis acceptance, as shown in Fig. 1, remains very close to 1. It increases with the 't Hooft coupling.

A first measure of efficiency is given by the average stepsize, i.e. the link change under each update. We measure this change simply by $\langle 1 - \frac{1}{N} \operatorname{Re} \operatorname{Tr} U_{\text{new}}^\dagger U_{\text{old}} \rangle$. The $SU(N)$ overrelaxation generates considerably larger steps than the $SU(2)$ subgroup approach, as visible in Fig. 3. The real test, of course, is the decorrelation of large Wilson loops. On our 2^4 lattice, we cannot probe large distances. Polyakov loops (Fig. 4, left) show critical slowing down as N increases, with a similar exponent

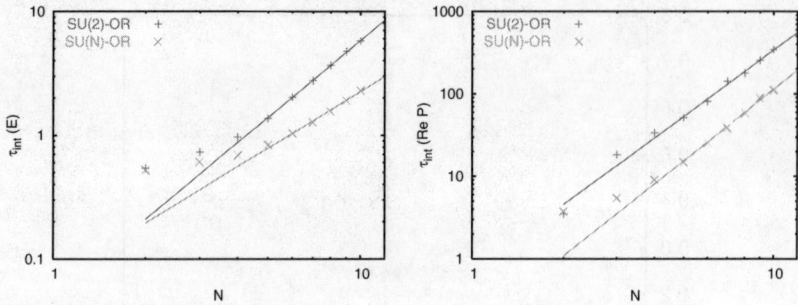

Fig. 4. Autocorrelation time of Polyakov loop (left) and of $(S_{\text{timelike}} - S_{\text{spacelike}})$ (right) versus N, for the two algorithms.

~ 2.8 using either update scheme. The $SU(N)$ strategy gives a speedup $\mathcal{O}(3)$, more or less independent of N. One observable, however, indicates a different dependence on N for the two algorithms. That is the asymmetry of the action, $\langle \sum_x \text{Re Tr}(\text{Plaq}^{\text{timelike}} - \text{Plaq}^{\text{spacelike}}) \rangle$. Fig. 4, right, shows that the speedup provided by the $SU(N)$ overrelaxation grows like $\sim N^{0.55}$. While this may be atypical, we never observed a slower decorrelation in the $SU(N)$ scheme for any observable.

In conclusion, overrelaxation in the full $SU(N)$ group appears superior to the standard $SU(2)$ subgroup approach. The results of [7] already indicated this for $SU(3)$. Our tests presented here suggest that the advantage grows with N, at least for some observables. For $SU(4)$ in (2+1) dimensions [18], the decorrelation of the Polyakov loop was ~ 3 times faster in CPU time, using $SU(N)$ overrelaxation, although our code implementation used simple calls to LAPACK routines, which are not optimized for operations on 4×4 matrices. We definitely recommend $SU(N)$ overrelaxation for large-N simulations.

Acknowledgements

Ph. de F. thanks the organizers for a stimulating conference, and Brian Pendleton in particular for his extraordinary patience with the proceedings. The hospitality of the Kavli Institute for Theoretical Physics, where this paper was written, is gratefully acknowledged.

References

1. Cabibbo, N. and Marinari, E.: "A new method for updating SU(N) matrices in computer simulations of gauge theories," Phys. Lett. B **119** (1982) 387.
2. Brown, F. R. and Woch, T. J.: "Overrelaxed heat bath and Metropolis algorithms for accelerating pure gauge Monte Carlo calculations," Phys. Rev. Lett. **58** (1987) 2394.

3. See, e.g., Lucini, B. and Teper, M.: "Confining strings in SU(N) gauge theories," Phys. Rev. D **64** (2001) 105019 [arXiv:hep-lat/0107007];
   Del Debbio, L., Panagopoulos, H., Rossi, P. and Vicari, E.: "Spectrum of confining strings in SU(N) gauge theories," JHEP **0201** (2002) 009 [arXiv:hep-th/0111090].
4. Olesen, P.: "Strings, tachyons and deconfinement," Phys. Lett. B **160** (1985) 408; see also Pisarski, R. D. and Alvarez, O.: "Strings at finite temperature and deconfinement," Phys. Rev. D **26** (1982) 3735.
5. Armoni, A., Shifman, M. and Veneziano, G.: "From super-Yang-Mills theory to QCD: planar equivalence and its implications," arXiv:hep-th/0403071.
6. Creutz, M.: "Overrelaxation and Monte Carlo simulation," Phys. Rev. D **36** (1987) 515.
7. Gupta, R., Kilcup, G. W., Patel, A., Sharpe, S. R. and de Forcrand, P.: "Comparison of update algorithms for pure gauge SU(3)," Mod. Phys. Lett. A **3** (1988) 1367.
8. Hasenbusch, M. and Necco, S.: "SU(3) lattice gauge theory with a mixed fundamental and adjoint plaquette action: lattice artefacts," JHEP **0408** (2004) 005 [arXiv:hep-lat/0405012].
9. Creutz, M.: "Monte Carlo study of quantized SU(2) gauge theory," Phys. Rev. D **21** (1980) 2308.
10. Kennedy, A. D. and Pendleton, B. J.: "Improved heat bath method for Monte Carlo calculations in lattice gauge theories," Phys. Lett. B **156** (1985) 393; Fabricius, K. and Haan, O.: "Heat bath method for the twisted Eguchi-Kawai model," Phys. Lett. B **143** (1984) 459.
11. Pietarinen, E.: "String tension in SU(3) lattice gauge theory," Nucl. Phys. B **190** (1981) 349.
12. Adler, S. L.: "An overrelaxation method for the Monte Carlo evaluation of the partition function for multiquadratic actions," Phys. Rev. D **23** (1981) 2901.
13. Heller, U. M. and Neuberger, H.: "Overrelaxation and mode coupling in sigma models," Phys. Rev. D **39** (1989) 616.
14. Decker, K. M. and de Forcrand, P.: "Pure SU(2) lattice gauge theory on 32**4 lattices," Nucl. Phys. Proc. Suppl. **17** (1990) 567.
15. Patel, A., Ph.D. thesis, California Institute of Technology, 1984 (unpublished); Gupta, R., Guralnik, G., Patel, A., Warnock, T. and Zemach, C.: "Monte Carlo renormalization group for SU(3) lattice gauge theory," Phys. Rev. Lett. **53** (1984) 1721.
16. Neuberger, H., private communication, and, e.g., Kiskis, J., Narayanan, R. and Neuberger, H.: "Does the crossover from perturbative to nonperturbative physics become a phase transition at infinite N?," Phys. Lett. B **574** (2003) 65. [arXiv:hep-lat/0308033].
17. Kennedy, A. D. and Bitar, K. M.: "An exact local hybrid Monte Carlo algorithm for gauge theories," Nucl. Phys. Proc. Suppl. **34** (1994) 786 [arXiv:hep-lat/9311017]; Horvath, I. and Kennedy, A. D.: "The local Hybrid Monte Carlo algorithm for free field theory: reexamining overrelaxation," Nucl. Phys. B **510** (1998) 367 [arXiv:hep-lat/9708024].
18. de Forcrand, P. and Jahn, O.: "Deconfinement transition in 2+1-dimensional SU(4) lattice gauge theory," Nucl. Phys. Proc. Suppl. **129** (2004) 709 [arXiv:hep-lat/0309153].

Improved Staggered Fermions

Eduardo Follana*

University of Glasgow, G12 8QQ Glasgow, UK. e.follana@physics.gla.ac.uk

Summary. At light quark masses, finite lattice spacing gives rise to spectrum doublers in the staggered fermion formalism, thus causing ambiguities in the extraction of physical quantities. We present results for the pion spectrum of simulations showing how improvements of the staggered fermion action remedy this deficiency.

1 Introduction

Accurate calculations of many Standard Model predictions require numerical simulations in lattice QCD with light dynamical quarks. The improved staggered formalism for light quarks is the only one capable of delivering this in the near future [1]. One problem of the staggered formalism is the presence of doublers in the spectrum, and of interactions between them, which we call taste-changing interactions. The existence of doublers implies that the spectrum of the theory is more complicated than the spectrum of QCD. Typically there are several staggered versions of a given QCD operator. They will all give the same answer in the continuum limit, when the lattice spacing vanishes, but the taste-changing interactions lift this degeneracy in a finite lattice, causing ambiguities in the extraction of physical quantities.

The improvement of the action is designed to decrease the strength of such interactions, and is crucial to obtain accurate physical predictions. It is therefore important to test with simulations the properties of such improved actions. Here we use the splittings in the pion spectrum, which are very sensitive to taste-breaking interactions, to test several variants of improved staggered actions.

2 From Naive to Staggered Fermions

Our starting point is the naive lattice discretization of quarks

---

*Work in collaboration with C. Davies (University of Glasgow), P. Lepage and Q. Mason (Cornell University), H. Trottier (Simon Fraser University), HPQCD and UKQCD collaborations.

$$S = \sum_x \bar{\psi}(x)(\gamma \cdot \Delta + m)\psi(x) \,.$$

$$\Delta_\mu \psi(x) = \frac{1}{2}(U_\mu(x)\psi(x+\hat{\mu}) - U_\mu^\dagger(x-\hat{\mu})\,\psi(x-\hat{\mu}))\,.$$

This naive discretization retains many important features of the QCD Lagrangian, such as a certain amount of chiral symmetry, and has small ($\mathcal{O}(a^2)$) discretization errors (a is the spacing of the lattice). It suffers, however, from the doubling problem. Each quark field actually describes 16 different species of particles (in 4 dimensions), which we will call tastes, to distinguish them from physical flavours. This can be understood as a consequence of an exact symmetry of the action, called doubling symmetry, which for a direction μ is given by

$$\psi(x) \to (i\gamma_5\gamma_\mu)\exp(i\pi x_\mu)\,\psi(x)\,.$$
$$\bar{\psi}(x) \to \bar{\psi}(x)(i\gamma_5\gamma_\mu)\exp(i\pi x_\mu)\,.$$

In general we have

$$\psi(x) \to \mathcal{B}_\zeta(x)\,\psi(x)\,. \tag{1}$$
$$\bar{\psi}(x) \to \bar{\psi}(x)\,\mathcal{B}_\zeta^\dagger(x)\,. \tag{2}$$

with

$$\mathcal{B}_\zeta(x) = \exp(i\pi\, x \cdot \zeta)\prod_\mu (i\gamma_5\gamma_\mu)^{\zeta_\mu}\,.$$
$$\zeta = (\zeta_1, \cdots, \zeta_4),\, \zeta_\mu \in \{0,1\}\,.$$

In order to define the staggered action, we use the fact that the naive quark action can be diagonalised in spin space by means of a unitary transformation

$$\psi(x) \to \Omega(x)\chi(x)\,.$$
$$\bar{\psi}(x) \to \bar{\chi}(x)\Omega^\dagger(x)\,.$$

where

$$\Omega(x) = \prod_\mu (\gamma_\mu)^{x_\mu}\,.$$

In the new variables χ, the action and the propagator are both diagonal in spinor space, for any gauge field

$$S = \sum_{x,\mu,\beta} \bar{\chi}_\beta(x)(\alpha_\mu(x)\Delta_\mu + m)\chi_\beta(x)\,. \tag{3}$$
$$<\chi(x)\bar{\chi}(y)> = g(x,y)\,I_{spin}\,.$$

with

$$\alpha_\mu(x) = (-1)^{x_1+\cdots+x_{\mu-1}}\,.$$

Equivalently, going back to the original variables, we see that the spinor structure of the naive quark propagator is independent of the gauge field,

$$<\psi(x)\bar{\psi}(y)> = g(x,y)\,\Omega(x)\Omega^\dagger(y)\,.$$

Here $g(x,y)$ contains all the dependence on the gauge field.

Therefore the naive quark formulation is redundant, and this allows us to "stagger" the quarks, and define the (naive) staggered action by keeping only one of the four spinor components in (3)

$$S_{\text{stagg}} = \sum_{x,\mu} \bar{\chi}(x)(\alpha_\mu(x)\Delta_\mu + m)\chi(x) \ .$$

This reduces the number of tastes from 16 down to 4.

3 Taste-Changing Interactions

The doubling transformations (1, 2) imply the existence of taste-changing interactions. A low energy quark that absorbs momentum close to $\zeta\pi/a$ is not driven far off-shell, but is rather turned into a low-energy quark of another taste. Thus the simplest process by which a quark changes taste is by the emission of a hard gluon of momentum $q \approx \zeta\pi/a$. This gluon is highly virtual and must be immediately reabsorbed by another quark whose taste will also change. Taste changes are therefore perturbative for typical values of the lattice spacing, and can be corrected systematically.

In order to test with numerical simulations the reduction of taste-changing interactions, we have calculated the spectrum of pions, where the effect of such interactions can be clearly seen. There are 16 different pions for staggered quarks, which would all be degenerate in the absence of taste-changing interactions. Due to taste-changing they split, and are grouped in 4 (approximate) multiplets. We will therefore use these splittings to gauge the improvement of the action.

4 Improved Actions

To remove taste-changing interactions at tree-level, a set of smeared paths can be adopted, in place of the original links [2, 3]. In Fig. 1, the first four paths are chosen to cancel the coupling between the quark field and a single gluon, A_μ, with momentum $\zeta\pi/a$, for any ζ. We will call such a smeared link the fat7 link[2]. It removes the high-momentum a^2 taste-changing interactions, but introduces an additional low-momentum a^2 error. This is corrected for by another 'staple' of links called the Lepage term. Conventionally this is discretised as the straight 5-link staple on the original link (the fifth path in Figure 1). A further a^2 error is removed by the 3-link Naik term (last path in Fig. 1). With these last two terms, plus tadpole improvement for all the links, the action is called asqtad improved staggered action.

It is possible to further smear the links in the action and obtain more improved actions. The HYP action [4] uses a 3-stage blocking procedure to smear the link in the derivative term. This action involves a step of projection to $SU(3)$ at each blocking stage.

The smearing process can be iterated, and we have also studied actions composed of two steps of fat7 or asqtad smearing, with or without projection onto $SU(3)$.

---

[2]The coefficients for the paths included in the fat7 action are different than those shown in Fig. 1, which are correct only for the asqtad action.

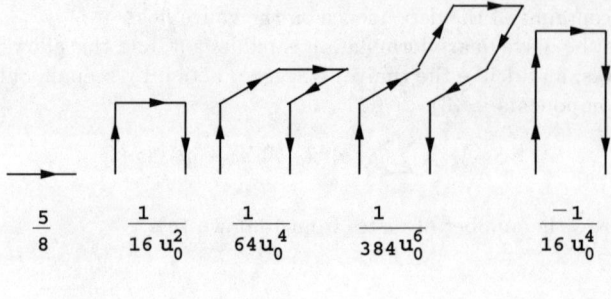

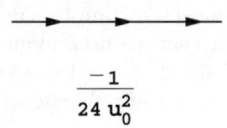

Fig. 1. Asqtad fat links

5 Pion Spectrum

The 16 different pions that can be constructed out of staggered fermions are generated, in naive language, by the set of operators

$$J_5(x) = \bar{\psi}(x)\,\gamma_5\,\psi(x) \, .$$
$$J_5^\mu(x) = \bar{\psi}(x+\hat{\mu})\,\gamma_5\,U_\mu^\dagger(x)\,\psi(x) \, .$$
$$J_5^{\mu\nu}(x) = \bar{\psi}(x+\hat{\mu}+\hat{\nu})\,\gamma_5\,U_\mu^\dagger(x+\hat{\nu})U_\nu^\dagger(x)\,\psi(x) \, .$$
$$J_5^{\mu\nu\rho}(x) = \bar{\psi}(x+\hat{\mu}+\hat{\nu}+\hat{\rho})\,\gamma_5\,U_\mu^\dagger U_\nu^\dagger U_\rho^\dagger\,\psi(x) \, .$$
$$J_5^{\mu\nu\rho\sigma}(x) = \bar{\psi}(x+\hat{\mu}+\hat{\nu}+\hat{\rho}+\hat{\sigma})\,\gamma_5\,U_\mu^\dagger U_\nu^\dagger U_\rho^\dagger U_\sigma^\dagger\,\psi(x) \, .$$

This operators are orthogonal to each other

$$< J_5^{(j)}(x) J_5^{(k)}(0) > \propto \delta_{jk} \, .$$

They are grouped into local, 1 link, 2 links, 3 links and 4 links multiplets, with degeneracy (1, 4, 6, 4, 1). The pions within each multiplet are approximately degenerate. The local one is the Goldstone boson. It is the only one that becomes massless at finite lattice spacing when the bare quark mass vanishes.

The pions would be degenerate in the absence of taste-changing interactions [3]. We will use the splittings between the multiplets to monitor the strength of such interactions. In the figures we show the difference between the mass squared of a given multiplet and the mass squared of the local pion. We only include in the plots the multiplets which do not involve links in the temporal direction, as they have smaller statistical errors.

Figure 2 shows the large improvement in going from the naive action to the asqtad action, as was already shown in previous studies [5, 6].

---

[3] We are discussing here only the "flavour nonsinglet" pions.

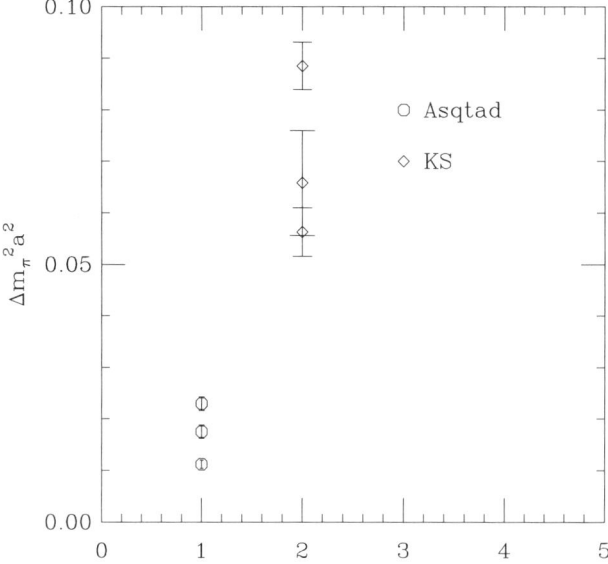

Fig. 2. Pion splittings for naive (KS) and asqtad actions.

In Fig. 3 we compare the results for several improved actions, along with perturbative estimates of the splittings. The first thing to notice is the further decrease in the splittings with more smearing, both for the HYP action and for the actions formed by iterated smearing. It is also interesting to notice that the introduction of a Lepage and Naik term (going from fat7 to asqtad) increases taste-breaking effects. Finally, the good agreement between the perturbative estimates and the simulation results should be noticed. It is apparent that perturbative calculations are a reliable guide to the effect of improvement.

In Fig. 4 we study the effect of projection onto $SU(3)$, by comparing the two best actions with and without projection. As can be clearly seen, the effect of the projection is considerable. The projection step can be difficult to incorporate into the usual simulation algorithms for dynamical quarks, and it is therefore important to explore alternatives.

6 Conclusions

Taste-changing interactions in the staggered quark formulation are a perturbative phenomenon, which can be removed systematically by improving the action. Perturbation theory is a reliable tool to choose the action and to estimate the size of taste-changing interaction effects. Going from the naive to the asqtad improved action reduces strongly the size of such effects, as seen in the pion spectrum. Further improvement is possible by increasing the amount of smearing. In such case, it is important to include steps that project the smeared links back onto $SU(3)$.

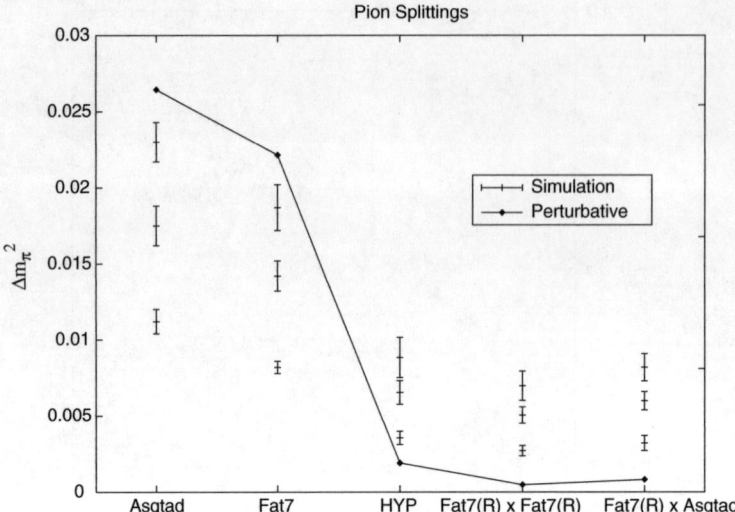

Fig. 3. Pion splittings in Simulation and Perturbation Theory. The R indicates reunitarization onto $SU(3)$.

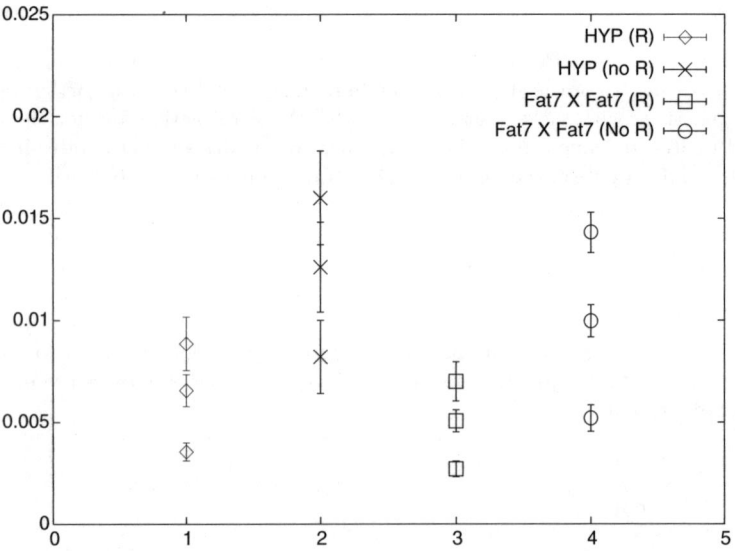

Fig. 4. Pion splittings with and without reunitarization. The R indicates reunitarization onto $SU(3)$.

References

1. K. Jansen, Nucl. Phys. B (Proc. Suppl.) **129&130** (2004) 3.
2. T. Blum *et al.*, Phys. Rev. D **55** (1997) 1133.
3. G. P. Lepage, Phys. Rev. D **59** (1999) 074502.
4. F. Knechtli, A. Hasenfratz, Phys. Rev. D63 (2001) 114502.
5. K. Orginos, D. Toussaint, R.L. Sugar, Phys. Rev. D **60** (1999) 054503.
6. E. Follana *et al.*, Nucl. Phys. B (Proc. Suppl.) **129&130** (2004) 384.

Perturbative Landau Gauge Mean Link Tadpole Improvement Factors

I.T. Drummond[1], A. Hart[2], R.R. Horgan[1], and L.C. Storoni[1]

[1] DAMTP, Cambridge University, Wilberforce Road, Cambridge CB3 0WA, U.K.
[2] School of Physics, University of Edinburgh, King's Buildings, Edinburgh EH9 3JZ, U.K.

Summary. We calculate the two loop Landau mean links for Wilson and improved SU(3) gauge actions, using twisted boundary conditions as a gauge invariant infrared regulator. We show that these numbers accurately describe high-β Monte Carlo simulations, and use these to infer the three loop coefficients.

1 Introduction

Tadpole improvement is now widespread in lattice field theory [1]. Without it, lattice perturbation theory begins to fail on distance scales of order 1/20 fm. Perturbation theory in other regularisations, however, seems to be phenomenologically successful down to energy scales of the order of 1 GeV (corresponding to lattice spacings of 0.6 fm) [2].

The reason is that the bare lattice coupling is too small [1, 2]. To describe quantities dominated by momenta of order the cut-off scale (π/a), it is appropriate to expand in the running coupling, α_s, evaluated at that scale. The bare coupling, however, deviates markedly from this, and its anomalously small value at finite lattice spacing can be associated with tadpole corrections [2]. These tadpole corrections are generally process independent, and can be (largely) removed from all quantities by modifying the action. This corresponds to a resumming of the perturbative series to yield an expansion in powers of a new, "boosted" coupling that is much closer to $\alpha_s(\pi/a)$.

Perturbatively this amounts to adding a series of radiative counterterms to the action. Such a series is obtained by dividing each gauge link in the action by an appropriate expansion, $u^{(\text{PT})}$. It is sufficient that this series is known only up to the loop order of the other quantities we wish to calculate using the action.

The factor $u^{(\text{PT})}$ is not unique, but it should clearly be dominated by ultraviolet fluctuations on the scale of the cut-off. The two most common definitions are the fourth root of the mean plaquette (a gauge invariant definition) and the expectation value of the link in Landau gauge. Both are successful, although there are some arguments for preferring the latter [3]. In this paper we discuss Landau gauge mean link tadpole improvement.

For Monte Carlo simulations, each gauge link in the action is divided by a numerical factor, $u^{(\mathrm{MC})}$. Its value is fixed by a self–consistency criterion; the value measured in the simulation should agree with the parameter in the action. Obtaining such numerical agreement requires computationally expensive tuning of the action, although linear map techniques can speed the convergence [4]. In some cases this cost is prohibitive, such as the zero temperature tuning of highly anisotropic actions for use in finite temperature simulations. As non–perturbative phenomena should not affect the cut-off scale, a sufficiently high order perturbative series should predict $u^{(\mathrm{MC})}$ such that the subsequent numerical tuning is unnecessary, or very slight.

In this paper we present the tadpole improvement factors calculated to two loop order using lattice perturbation theory. This covers the loop order of most perturbative calculations using lattice actions. In addition, we perform Monte Carlo simulations over a range of gauge couplings extending from the high-β regime down to the lattice spacings used in typical simulations. We show that the two loop predictions agree very well (showing finite volume effects are not significant to the tadpole improvement factors. For this reason we refer to both $u^{(\mathrm{PT})}$ and $u^{(\mathrm{MC})}$ as u from hereonin). The small deviations at physical couplings are shown to be consistent with a third order correction to u, and we infer the coefficient of this [5].

We demonstrate that the two loop formula predicts the numerically self–consistent $u^{(\mathrm{MC})}$ to within a few digits of the fourth decimal place, and the additional tuning required is minimal (especially when the action can be rescaled as in [6, 7]). In most cases no tuning is required if the third order is included.

These calculations are carried out for two SU(3) lattice gauge actions; the Wilson action and a first order Symanzik improved action. Isotropic and anisotropic lattices are studied, and interpolations in the anisotropy are given.

The results presented here extend and, in some cases, correct the preliminary results presented in [8] and are more fully described in [9]. Extension of this work to actions including fermions is being carried out, and will be reported in a future publication.

1.1 The Actions

We consider Wilson (W) and the Symanzik improved (SI [10]) actions:

$$S_W = -\frac{\beta_0}{\chi_0} \sum_{x,s>s'} \frac{P_{s,s'}}{u_s^4} - \beta_0 \chi_0 \sum_{x,s} \frac{P_{s,t}}{u_s^2 u_t^2} ,$$

$$S_{SI} = -\frac{\beta_0}{\chi_0} \sum_{x,s>s'} \left(\frac{5}{3} \frac{P_{s,s'}}{u_s^4} - \frac{1}{12} \frac{R_{ss,s'}}{u_s^6} - \frac{1}{12} \frac{R_{s's',s}}{u_s^6} \right)$$

$$- \beta_0 \chi_0 \sum_{x,s} \left(\frac{4}{3} \frac{P_{s,t}}{u_s^2 u_t^2} - \frac{1}{12} \frac{R_{ss,t}}{u_s^4 u_t^2} \right) , \qquad (1)$$

Table 1. Fits of infinite volume extrapolated perturbative coefficients to functions of χ.

action	quantity	const.	$1/\chi$	$1/\chi^2$	$\log\chi/\chi$
Wilson	$a_s^{(1)} = b_s^{(1)}$	-0.1206	0.1031	-0.0600	-0.0614
	$a_t^{(1)} = b_t^{(1)}$	0	-0.0169	-0.0609	0.0088
	$a_s^{(2)}$	-0.0174	0.0199	-0.0235	-0.0411
	$a_t^{(2)}$	0	0.0038	-0.0250	-0.0012
	$b_s^{(2)}$	0.0296	-0.0415	0.0149	-0.0134
	$b_t^{(2)}$	0	0.0180	-0.0153	-0.0085
Symanzik	$a_s^{(1)} = b_s^{(1)}$	-0.1004	0.0916	-0.0542	-0.0537
improved	$a_t^{(1)} = b_t^{(1)}$	0	-0.0123	-0.0527	0.0066
	$a_s^{(2)}$	-0.0300	-0.0215	0.0383	0.0373
	$a_t^{(2)}$	0	0.0071	-0.0214	-0.0042
	$b_s^{(2)}$	0.0012	-0.0728	0.0743	0.0645
	$b_t^{(2)}$	0	0.0129	-0.0109	-0.0068

where s, s' run over spatial links, $P_{s,s'}$ and $P_{s,t}$ are 1×1 plaquettes, $R_{ss,s'}$ and $R_{ss,t}$ are 2×1 loops. The bare gauge coupling is $g_0^2 = 6/\beta$ and χ_0 is the bare anisotropy.

We may minimise the dependence of the action on the self–consistent tadpole improvement factors through a rescaling

$$\beta = \frac{\beta_0}{u_s^3 u_t} \equiv \frac{6}{g^2}, \quad \chi = \frac{\chi_0 u_s}{u_t}. \qquad (2)$$

This removes all tadpole factors in the Wilson action. In the SI case, the residual factors are expanded perturbatively, with the radiative terms treated as counterterms in the action. This is convenient for perturbation theory, but

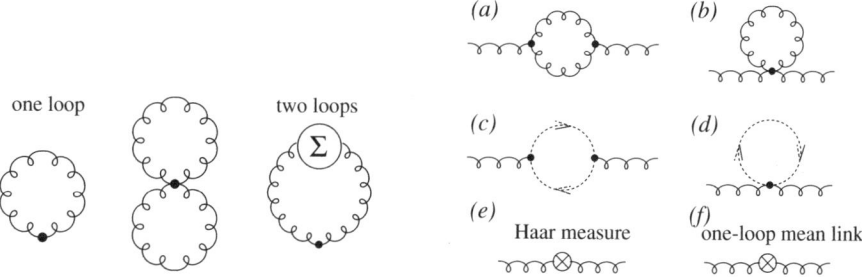

Fig. 1. Feynman diagrams for the Landau mean link. The contributions (a-f) to the self energy, Σ, are shown on the right–hand side.

also is useful when, for instance, self–consistently tuning action parameters in Monte Carlo simulations [7].

If $u_s = 1 + d_s g^2 + O(g^4)$, then

$$S_{SI}(\beta,\chi,u_s) = S_{SI}(\beta,\chi,u_s = 1) + g^2 \Delta S_{SI}$$
$$\Delta S_{SI} = \tfrac{\beta d_s}{6}\left(\chi R_{ss,t} + \chi^{-1}(R_{ss,s'} + R_{s's',s})\right)$$

and $g^2 \Delta S_{SI}$ becomes an insertion in the gluon propagator (Fig. 1(f)).

2 Landau Mean Link Improvement

The quantity used for tadpole improvement should be dominated by the short distance fluctuations of the action. To leading order, the gauge link is unity with all deviations arising from tadpole contributions. We focus here on the Landau mean link scheme, where $u_{s,t}$ are defined as the expectation values of spatial and temporal links in Landau gauge.

The perturbative series for $u_{l=s,t}$ can be written as

$$u_l \equiv \frac{1}{3}\langle \mathrm{Tr}\, U_l\rangle = \begin{cases} 1 + a_l^{(1)} g^2 + a_l^{(2)} g^4 + O(g^6), \\ 1 + b_l^{(1)} g_0^2 + b_l^{(2)} g_0^4 + O(g_0^6). \end{cases} \qquad (3)$$

The "bare" coefficients $a_l^{(i)}$ were calculated, and the series then re–expressed using the "boosted" coupling, $g_0{}^2$, as per Eqn. (2):

$$b_l^{(1)} = a_l^{(1)},$$
$$b_l^{(2)} = a_l^{(2)} + a_l^{(1)}\left(3a_s^{(1)} + a_t^{(1)}\right). \qquad (4)$$

The Feynman rules are obtained by perturbatively expanding the lattice actions using an automated computer code written in PYTHON [11, 6, 8, 12].

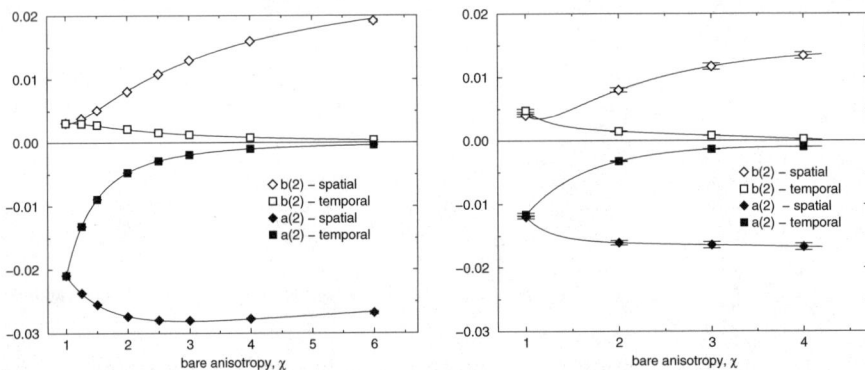

Fig. 2. Two loop Landau mean link coefficients for the Wilson (left–hand side) and SI actions.

The additional vertices associated with the ghost fields and the measure are given in [11, 6]. The gluon propagator in Landau gauge is discussed in [6, 8]. The Feynman diagrams are shown in Fig. 1.

We follow [11] and use twisted periodic boundary conditions for the gauge field. There is no zero mode and no concomitant infrared divergences whilst gauge invariance is still maintained. All choices of boundary condition are equivalent in the infinite volume limit. For the perturbation theory we employ doubly twisted boundary conditions [13, 6].

The one loop integration is carried out by summation of all twisted momentum modes for lattices $16 \leq L_\mu \leq 32$ on a workstation. To speed up the approach to infinite volume, the momenta are "squashed" in the directions with periodic boundary conditions using the change of variables $\mathbf{k} \to \mathbf{k}'$:

$$k'_\mu = k_\mu - \alpha_\mu \sin(k_\mu), \tag{5}$$

giving an integrand with much broader peaks [11]. Choosing $\alpha_\mu \sim 1 - (\chi L_\mu)^{-1}$ significantly reduces the dependence on L. All results were extrapolated to infinite volume using a fit function of the form $c_0 + c_1/L^2$.

The two loop calculations used a parallel version of VEGAS, a Monte Carlo estimation program [14, 15], on between 64 and 256 processors of a Hitachi SR2201 supercomputer. Each run took of the order of 24 hours.

We interpolate the perturbative coefficients using the function $c_0 + c_1/\chi + c_2/\chi^2 + c_3 \log \chi / \chi$. The fit parameters are given in Table 1 and are shown in Fig. 2. We stress that χ is the anisotropy after the majority of tadpole factors have been scaled out of the action.

3 Comparison with Monte Carlo Simulation Results

By comparing such truncated series with results from high-β Monte Carlo simulations, we may check the accuracy of our calculations, the effect of finite simulation volumes and infer higher order coefficients in the perturbative expansion [16]. Field configurations were generated using a 2nd order Runge-Kutta Langevin updating routine. The Langevin evolution algorithm is coded such that any pure gauge action can be simulated by simply specifying the list of paths and the associated couplings. The group derivative for each loop is then computed automatically by an iterative procedure which moves around the loop link by link constructing the appropriate traceless anti-Hermitian contribution to the Langevin velocity. This is the most efficient implementation, minimising the number of group multiplications needed. It applies whenever the quantity to be differentiated is specified as a Wilson path.

The twisted boundary conditions are implemented in the manner suggested in [11] where the field simulated, $\bar{U} = U$ everywhere, save $U_\mu(\mathbf{x})\Omega_\mu$ when $x_\mu = L_\mu$, where Ω_μ is the twist matrix associated with the μ direction and L_μ is the lattice extent. The action $S(\bar{U})$ is identical to the untwisted action

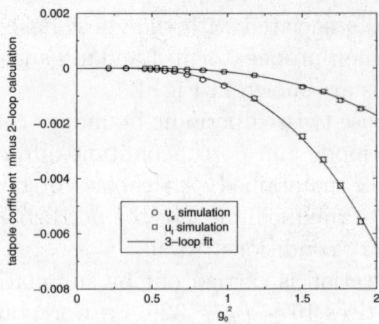

Fig. 3. Deviation, u_-, of the Landau mean link from two loop PT as a function of g_0^2 for the Symanzik-improved action with $\chi = 2$. The solid lines show fits with 3 loop $O(g_0^6)$ parametrization.

except that a loop whose projection onto the (μ, ν)-plane, $\mu < \nu$, encircles the point $(L_\mu + \frac{1}{2}, L_\nu + \frac{1}{2})$ has an additional factor of $(z_{\mu\nu})^{-c}$ where c is the integer winding number. The program assigns the correct phase factors to each path appropriately. A Fourier accelerated algorithm is vital when fixing to Landau gauge, which is otherwise prohibitively time consuming. We found the convergence of the gauge fixing to be fastest for quadruply twisted boundary conditions when studying anisotropic lattices.

The mean link was measured for the Symanzik improved action for a range of couplings $5.2 \leq \beta \leq 30$. The lower values are within to the physical region: as a guide, $\beta_0 = 2.4$ at $\chi_0 = 3$ in [17] corresponds to $\beta = 6.37$ and a lattice spacing of $a = 0.16$ fm. The lattices used had $\chi = 2$ and size $L^3 T = 8^3 16$, giving a hypercubic volume with equal sides in physical units. To show the deviation of the results from perturbation theory, in Fig. 3 we plot the simulation data with the two loop perturbative prediction already subtracted.

As discussed above, the simulation boundary conditions were twisted in all four directions. To compare with the perturbative calculation an adjustment for finite L and T effects is therefore needed. We computed the one loop tadpole coefficients for the appropriate finite-size values but this proved more difficult in the two loop case since we applied a twist in two directions only and achieved most rapid convergence in the integrals by using the change of variables in Eqn. (5) which is unavailable in the simulation. Of course, the $L, T \to \infty$ result is independent of the precise details but for the comparison with simulation we must allow for a small discrepancy with our theory. The results are shown in in Fig. 3 and we concentrate on the results for the Symanzik-improved action. Both the spatial (u_s) and temporal (u_t) tadpole coefficients are plotted and the fit lines shown are of the form

$$u_- = \delta b_i^{(2)} g_0^4 + b_i^{(3)} g_0^6 \, ,$$

where $\delta b_i^{(2)}$ is the finite-size correction at two loops. Fitting the full range of data, the $\chi^2/$d.o.f. is around 0.6 and the fit coefficients are

$$\delta b_s^{(2)} = 0.00032(4), \quad b_s^{(3)} = -0.00097(3),$$
$$\delta b_t^{(2)} = 0.00015(1), \quad b_t^{(3)} = -0.00029(1). \tag{6}$$

The finite-size correction corresponds to a 2% correction due to finite-size effects in the two loop calculation. This is certainly very reasonable. The measurement of the three loop coefficients is accurate to 3% and we should also expect a finite-size error of a similar order. The main conclusion is that even in the physical region $\beta \sim 5-6$ perturbation theory works very well indeed for the Landau mean-link and, moreover, there is no observable deviation from a three loop $O(g_0^6)$ perturbative approximation. This is very encouraging for the accurate design of QCD actions and the corresponding perturbative analyses based upon them.

We are pleased to acknowledge the use of the Hitachi SR2201 at the University of Tokyo Computing Centre and the Cambridge–Cranfield High Performance Computing Facility. We thank N.A. Goodman and A.J. Craig for their contribution. A.H. is supported by the Royal Society.

References

1. G. P. Lepage and P. B. Mackenzie, Phys. Rev. **D48**, 2250 (1993), [hep-lat/9209022].
2. G. P. Lepage, hep-lat/9607076.
3. P. Lepage, Nucl. Phys. Proc. Suppl. **60A**, 267 (1998), [hep-lat/9707026].
4. M. G. Alford, I. T. Drummond, R. R. Horgan, H. Shanahan and M. J. Peardon, Phys. Rev. **D63**, 074501 (2001), [hep-lat/0003019].
5. H. D. Trottier, N. H. Shakespeare, G. P. Lepage and P. B. Mackenzie, Phys. Rev. **D65**, 094502 (2002), [hep-lat/0111028].
6. I. T. Drummond, A. Hart, R. R. Horgan and L. C. Storoni, Phys. Rev. **D66**, 094509 (2002), [hep-lat/0208010].
7. I. T. Drummond, A. Hart, R. R. Horgan and L. C. Storoni, Phys. Rev. **D68**, 057501 (2003), [hep-lat/0307010].
8. I. T. Drummond, A. Hart, R. R. Horgan and L. C. Storoni, Nucl. Phys. Proc. Suppl. **119**, 470 (2003), [hep-lat/0209130].
9. A. Hart, R. R. Horgan and L. C. Storoni, Phys. Rev. **D70**, 034501 (2004), [hep-lat/0402033].
10. M. Alford, T. Klassen and G. Lepage, Phys. Rev. **D58**, 034503 (1998).
11. M. Lüscher and P. Weisz, Nucl. Phys. **B266**, 309 (1986).
12. A. Hart, G. M. von Hippel, R. R. Horgan and L. C. Storoni, hep-lat/0411026.
13. A. Gonzalez-Arroyo, hep-th/9807108.
14. G. P. Lepage, J. Comp. Phys. **27**, 192 (1978).
15. W. Press *et al.*, Numerical Recipes: the art of scientific computing (2nd ed., CUP, 1992).
16. H. D. Trottier *et al.*, Phys. Rev. **D65**, 094502 (2002), [hep-lat/0111028].
17. J. Shigemitsu *et al.*, Phys. Rev. **D66**, 074506 (2002), [hep-lat/0207011].

Reversibility and Instabilities in Hybrid Monte Carlo Simulations

Bálint Joó

School of Physics, University of Edinburgh, Edinburgh EH9 3JZ, Scotland, United Kingdom, bj@ph.ed.ac.uk

Summary. It has long been known in the lattice community, that molecular dynamics (MD) integrators used in lattice QCD simulations can suffer from instabilities. In this contribution we review pedagogically where these instabilities arise, how they may be noticed and what actions can be taken to avoid them. The discussion is of relevance to simulations with light quarks such as those attainable using Ginsparg Wilson fermions.

1 Introduction

Of the currently available algorithms for generating gauge configurations in simulations of lattice QCD with dynamical fermions, the Hybrid Monte Carlo (HMC) algorithm [1] in one of its many guises is of the most common use.

A feature of HMC like algorithms is their reliance on molecular dynamics (MD) integrators. It has been noticed in [3] and [4] that the typical leap-frog integrator can show chaotic behaviour in lattice QCD simulations and that the integrator can go unstable. Further investigation [5, 6] has shown that the instability in the integrator can manifest itself on production size lattices at realistic parameters. In this contribution we will present our current understanding of instabilities in MD integrators used in the HMC algorithm.

The remainder of this contribution is organised as follows. In section 2 we review the relevant features of the HMC algorithm; in section 3 we show how instabilities can arise in the MD part of the HMC algorithm by referring to the case of a simple harmonic oscillator; in section 4 we will relate this to QCD data; and we draw our conclusions in section 5.

2 HMC and Molecular Dynamics

HMC simulations generally proceed through the following steps:

i) *Momentum Refreshment:* new momenta are generated from a Gaussian heatbath.
ii) *Molecular Dynamics:* the canonical coordinates and momenta are evolved by solving the equations of motion numerically.

iii) *Accept/Reject step:* the new coordinates and momenta at the end of the MD step are subjected to an accept/reject test. This is typically either Metropolis [7] like or a Noisy Linear Accept/Reject step [8]. If the new coordinates and momenta are rejected the new phase space state is taken to be the initial coordinates and momenta from before the MD.

For further details of these steps we suggest the reader consult [2] in these proceedings. In this article we will concentrate mostly on the MD step.

For the HMC algorithm to be convergent, it must be ergodic and have the desired equilibrium distribution as its fixed point. Ergodicity is (hopefully) provided by the momentum refreshment (which boosts the system to a different randomly chosen energy hyperplane). For the fixed point condition to be satisfied it is sufficient that the algorithm satisfies detailed balance. Let us suppose that we are trying to generate coordinates q and momenta p according to some equilibrium probability distribution $P_{eq}(q,p)$, and that in the above procedure, the MD generates new coordinates and momenta (q',p') from an initial state of (q,p) with probability $P(q',p' \leftarrow q,p)$. The detailed balance condition can be written as:

$$P(q',p' \leftarrow q,p) P_{eq}(q,p) \, dq \, dp = P(q,p \leftarrow q',p') P_{eq}(q',p') \, dq' dp'$$

In order to satisfy detailed balance, it is necessary for the MD step to be reversible and measure (area) preserving. Typically, a symmetric integrator constructed from symplectic update components is used. The symplectic components guarantee area preservation and their symmetric combination ensures the reversibility.

Let us consider symplectic update operators $\mathcal{U}_q(\delta\tau)$ and $\mathcal{U}_p(\delta\tau)$ acting on a phase space state (q,p) as:

$$\mathcal{U}_q(\delta\tau)\,(q,p) = (q + p\delta\tau,\; p)$$
$$\mathcal{U}_p(\delta\tau)\,(q,p) = (q,\; p + F\delta\tau)$$

where F is the MD Force term, computed from the action $S(q)$ as

$$F = -\frac{\partial S}{\partial q}.$$

Both $\mathcal{U}_q(\delta\tau)$ and $\mathcal{U}_p(\delta\tau)$ are measure preserving and for reversibility one needs to use a symmetric combination such as the PQP leap-frog integrator:

$$\mathcal{U}_3(\delta\tau) = \mathcal{U}_p\left(\frac{\delta\tau}{2}\right) \mathcal{U}_q(\delta\tau) \mathcal{U}_p\left(\frac{\delta\tau}{2}\right) \tag{1}$$

which is of typical use in lattice QCD simulations and has errors of $O(\delta\tau^3)$ per time step.

The construction of higher order integration schemes is possible. In this work we consider the schemes of [9, 10]. If an integrator defined by update operator $\mathcal{U}_{n+1}$ is available which is accurate to $O(\delta\tau^{n+1})$ for some even natural number n, one can construct an integrator accurate to $O(\delta\tau^{n+3})$ by combining the $\mathcal{U}_{n+1}$ integrators as:

$$\mathcal{U}_{n+3}(\delta\tau) = \mathcal{U}_{n+1}(\delta\tau_1)^i \, \mathcal{U}_{n+1}(\delta\tau_2) \, \mathcal{U}_{n+1}(\delta\tau_1)^i \tag{2}$$

$$\text{with } \delta\tau_1 = \frac{\delta\tau}{2i - s}$$
$$\delta\tau_2 = -s\delta\tau_1$$
$$\text{and } s = (2i)^{\frac{1}{n+2}} \tag{3}$$

for some arbitrary positive integer i.

Finally we consider another class of symplectic integrators due to Sexton and Weingarten [11] which rely on splitting the action $S(q)$ into parts that are easy and hard to integrate respectively. Writing

$$S(q) = S_1(q) + S_2(q)$$

we consider update operators $\mathcal{U}_1$ and $\mathcal{U}_2$, which are now *not* improved in the sense of eqs. (2)-(3). Instead, the subscript labels signify that they update the coordinates and momenta according to the forces defined from $S_1(q)$ and $S_2(q)$ respectively. Since the two improvement schemes are orthogonal to each other, the meaning of the subscripts should always be clear from the context.

A full update operator which updates the system according to $S(q)$ can then be constructed as:

$$\mathcal{U}_S(\delta\tau) = \mathcal{U}_2\left(\frac{\delta\tau}{2}\right)\left[\mathcal{U}_1\left(\frac{\delta\tau}{n}\right)\right]^n \mathcal{U}_2\left(\frac{\delta\tau}{2}\right) \qquad n > 0 \ .$$

We note that if we identify $\mathcal{U}_1$ with $\mathcal{U}_q$ and $\mathcal{U}_2$ with $\mathcal{U}_p$ and choose $n = 1$, then we recover in $\mathcal{U}_S$ the 3rd order PQP leap-frog $\mathcal{U}_3$ of (1). Integrators of this type are accurate to $O(\delta\tau^3)$ per time step, however, now the coefficient of the third order error term decreases as $\frac{1}{n^2}$. The expected gain from this scheme is that the step–size error can be reduced by repeated application of computationally cheaper updates $\mathcal{U}_1$ while keeping the same number of applications of $\mathcal{U}_2$.

3 Instability in the Simple Harmonic Oscillator (SHO)

We now apply the various integration schemes to the SHO, and demonstrate the onset of an instability in this simplest of systems. We consider an SHO with action

$$S(q) = \frac{1}{2}\omega^2 q^2$$

and Hamiltonian

$$H(q,p) = \frac{1}{2}p^2 + S(q) \ .$$

The corresponding force term is

$$F = -\frac{\partial S}{\partial q} = -\omega^2 q$$

which is linear in q. Consequently we can write the PQP leap-frog update of the system as a matrix equation:

$$\begin{pmatrix} q(t+\delta\tau) \\ p(t+\delta\tau) \end{pmatrix} = \begin{pmatrix} 1 - \frac{1}{2}(\omega\delta\tau)^2 & \omega\delta\tau \\ -\omega\delta\tau + \frac{1}{4}(\omega\delta\tau)^3 & 1 - \frac{1}{2}(\omega\delta\tau)^2 \end{pmatrix} \begin{pmatrix} q(t) \\ p(t) \end{pmatrix} \ .$$

The eigenvalues of the update matrix above are:

$$\lambda_\pm = 1 - \frac{1}{2}(\omega\delta\tau)^2 \pm \sqrt{1 - \frac{1}{4}(\omega\delta\tau)^2} = e^{\pm i \cos^{-1}\left(1 - \frac{1}{2}(\omega\delta\tau)^2\right)} \ .$$

One can see at once that for $\omega\delta\tau < 2$ the eigenvalues $\lambda_\pm$ are imaginary and rotate in the complex plane; however for $\omega\delta\tau > 2$ the arguments of the $\cos^{-1}$ term become purely imaginary resulting in eigenvalues which are real and of which one is exponentially increasing with $\delta\tau$ driving the system unstable. At the same time, the phase space orbits of the oscillator system change from being elliptic to hyperbolic.

To see the idea in a more general way consider the update operator of a symplectic integrator $\mathcal{U}$. Since it is area preserving we have $\det \mathcal{U} = 1$ and since all the components of $\mathcal{U}$ are real, the trace $\mathrm{Tr}\,\mathcal{U}$ is also real. Writing the two eigenvalues of $\mathcal{U}$ as

$$\lambda_1 = u_1 + iv_1 \quad \text{and} \quad \lambda_2 = u_2 + iv_2$$

we have

$$v_1 = -v_2 \quad \text{and} \quad u_1 v_2 + u_2 v_1 = 0 \;.$$

The conditions on the determinant and the trace then leave two possibilities:

Stable Region

$$u_1 = u_2 \Rightarrow u_1^2 + v_1^2 = 1 \Rightarrow \lambda_{1,2} = e^{\pm i\theta}$$

for some real θ.

Unstable Region

$$v_1 = v_2 = 0 \Rightarrow \lambda_1 = \eta, \lambda_2 = \frac{1}{\eta}$$

for some $\eta \geq 1$.

The general feature is that the instability occurs when eigenvalues change from imaginary to real, or in other words when the discriminant of the update matrix changes from positive to negative. Using this observation we can show that the 5th order system of [9, 10] should be unstable when $\omega\delta\tau \in \{0, \sqrt{12 - 6\sqrt[3]{4}}\}$ and the 7th order integrator goes unstable when $\omega\delta\tau \in (1.592, 1.822)$ and $\omega\delta\tau > 1.869$. Note that in both the higher order schemes the situation is worse than in the 3rd order leap-frog scheme as in the higher order schemes the instabilities occur at smaller values of $\omega\delta\tau$.

As for the scheme of [11], if $\mathcal{U}_1 = \mathcal{U}_\mathrm{q}$ and $\mathcal{U}_2 = \mathcal{U}_\mathrm{p}$ then we would expect the the integrator $\mathcal{U}_\mathrm{S}$ to go unstable at the same value of $\omega\delta\tau$ as the normal PQP integrator, seeing that the term $\omega\delta\tau$ appears only in the $\mathcal{U}_\mathrm{p} = \mathcal{U}_2$ step which has the same step size $\frac{\delta\tau}{2}$ in both schemes.

4 Instabilities and QCD Simulations

Instabilities in lattice QCD simulations were originally reported by [3, 4] on very small system sizes (consisting of $V = 4^4$ lattice sites.) The authors also reported observing chaos in the numerical solution of the MD equations of motion. While the contribution of [4] and the later papers of [3] were simultaneous and are equally important, our own work in this area is a clear follow up to the work reported in [4]. Hence we will concentrate mostly on those results and our own below. However we do urge the reader to also keep in mind the results of [3].

The issue of chaos in the equations of motion is related to the reversibility of the numerical implementation of the MD integration scheme. As mentioned in section 2, reversibility is generally required for the detailed balance condition to hold. While the schemes discussed in sections 2 and 3 are explicitly reversible in exact arithmetic, a floating point implementation can suffer from rounding errors. If the integrator is chaotic, these rounding errors are magnified exponentially with a characteristic exponent generally referred to as the (first) Lyapunov exponent.

In [4] and also in [3] it was shown that indeed the MD integrator is chaotic (with a positive Lyapunov exponent). However, metrics of reversibility including the Lyapunov exponent itself seem not to grow with step size $\delta\tau$ until some critical value which the authors of [4] ascribed to be the instability in the integrator. Once the instability was encountered, reversibility metrics became very large showing a distinct lack of reversibility in the integrator. However, this did not present a problem, since at the critical and larger values of $\delta\tau$ the energy change along a trajectory δH was already so large that the acceptance test at the end of the trajectory would never succeed when the integrator was used in an HMC context. It was also noticed that for lighter quarks this growth of the Lyapunov exponent set in at smaller step-sizes $\delta\tau$ than for heavier quarks or for quenched (infinitely heavy quarks) QCD.

The authors of [4] made the hypothesis that *since an asymptotically free field theory like QCD essentially acts like a loosely coupled system of harmonic oscillator modes, there will be a few high frequency modes which drive the system unstable at a suitably large step size. The lower frequency bulk modes can smooth the transition from stability to instability, so that it does not occur as abruptly as it does for just a single mode.* The frequency of the modes can be characterised by the force vector. The infinity norm of the force $||F||_\infty$ selects the largest element of the force vector and this characterises the highest frequency mode. The 2-norm of the Force $||F||_2$ divided by the number of degrees of freedom $||F||_2/\text{d.o.f}$ characterises the size of the average bulk mode. It was conjectured that the instability sets in when the $F\delta\tau$ term in the momentum update step of the integrator reaches a critical value (akin to the $\omega\delta\tau$ term in the harmonic oscillator system). Furthermore the maximum frequency of the modes should grow in proportion to some inverse power of the quark mass (on dimensional grounds) which serves to explain why the instability sets in at smaller step sizes for lighter quark systems.

In [5, 6] we measured the norms of the forces along each step of an MD trajectory for a variety of mass and step size values. Our measurements were made on relatively large lattices of size $V = 16^3 \times 32$ sites, using step sizes and mass parameters (κ) that were typical of a large scale HMC simulation at the time.

We show how the fermionic force varies with quark mass in figure 1. We fit data with the model: $F = C(am)^\alpha$, leaving C and α as free parameters. Our quark mass am is defined as

$$am = \frac{1}{2}\left(\frac{1}{\kappa} - \frac{1}{\kappa_c}\right)$$

where κ_c is the critical value of kappa corresponding to massless quarks. For the purposes of determining am for the points, we used a value of κ_c which was determined independently in [12] to be $\kappa_c = 0.13633$. However, for the purposes of fitting, we left κ_c as a free parameter in our fits and were gratified when the fit results were consistent with our independent determination. We note that the fits measure a small but negative value of α indicating that the force norms do indeed vary with some inverse power of the quark mass which is entirely consistent with the hypothesis outlined above.

In figure 2 we show the force norms, and the change in energy along a trajectory (δH) for a variety of step sizes $\delta\tau$ at a fixed value of the mass parameter κ. We

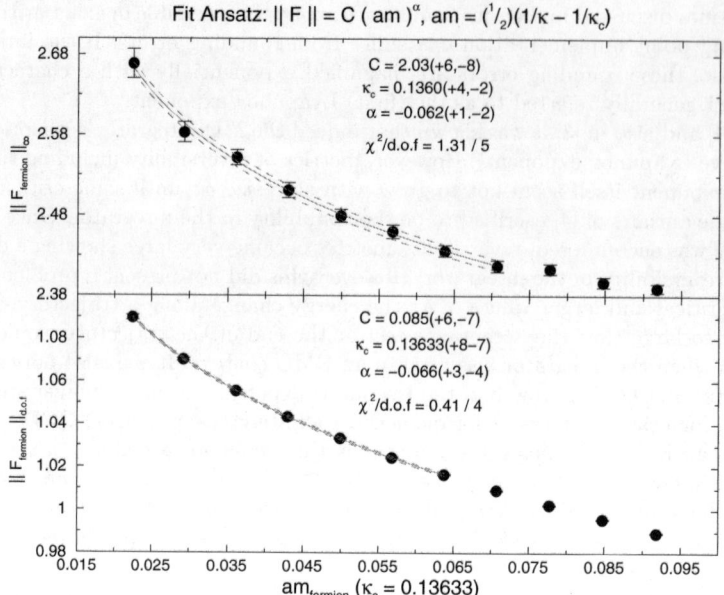

Fig. 1. The force norms against the quark mass am for a particular value of $\delta\tau$.

separate the force into gauge and fermionic contributions. It can be seen that the fermionic force norms start to increase at $\delta\tau = 0.011$ accompanied by a sudden increase in δH. Typically δH needs to be $O(1)$ for reasonable acceptance rates. Here it has jumped from $O(1)$ to $O(100)$ as the critical value of $\delta\tau$ was crossed.

In figure 3 we show the ∞-norms of the gauge and fermionic forces and the energy change δH against the mass parameter κ at two values of $\delta\tau$, one for which the system remains stable and one for which it goes unstable. We can see that as the quarks become lighter (κ becomes larger) we reach a value of κ where the fermionic force suddenly grows accompanied by a large jump in δH at $\delta\tau = 0.012$. This does not happen for the smaller step size of $\delta\tau = 0.01$.

This data all seems entirely consistent with the earlier hypothesis. The integrator goes unstable at a critical combination of the force and the step size. The instability is manifested by a large increase in the energy change δH along a trajectory which then results in drop in the acceptance rate of the HMC algorithm. If the quarks are sufficiently light, this can happen at values of $\delta\tau$ typical of a production run.

Finally, we note that in dynamical fermion simulations the fermionic force contains the result of solving a large sparse system of equations, usually using a Krylov subspace method. There is no requirement in the HMC algorithm to solve this system exactly or even to machine precision, as long as the solution procedure is reversible, which can be achieved by always using the same initial vector for the solution or by using a random initial vector. Starting the solution with the result vector from the last MD step is not reversible unless the solution is carried out with such high accuracy as to be essentially exact (and independent of the starting vector). Since generally a large fraction of computer time is spent performing this solution process,

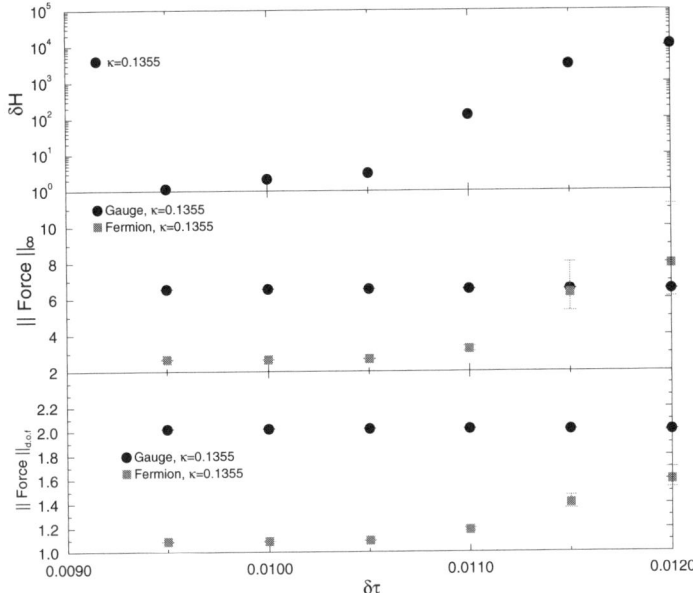

Fig. 2. The force norms and the energy change δH along a trajectory, against the step size $\delta\tau$ for $\kappa = 0.1550$.

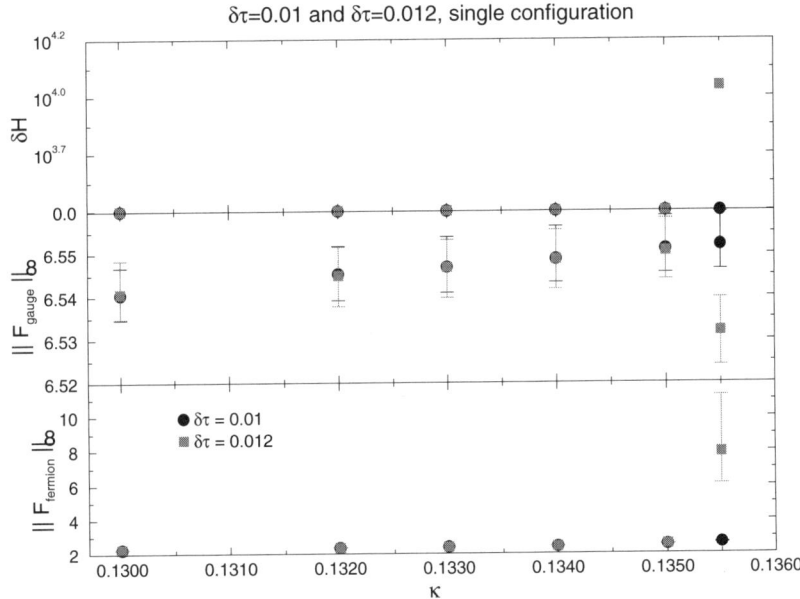

Fig. 3. The force norms and the energy change δH along a trajectory, against the mass parameter κ for a fixed $\delta\tau$.

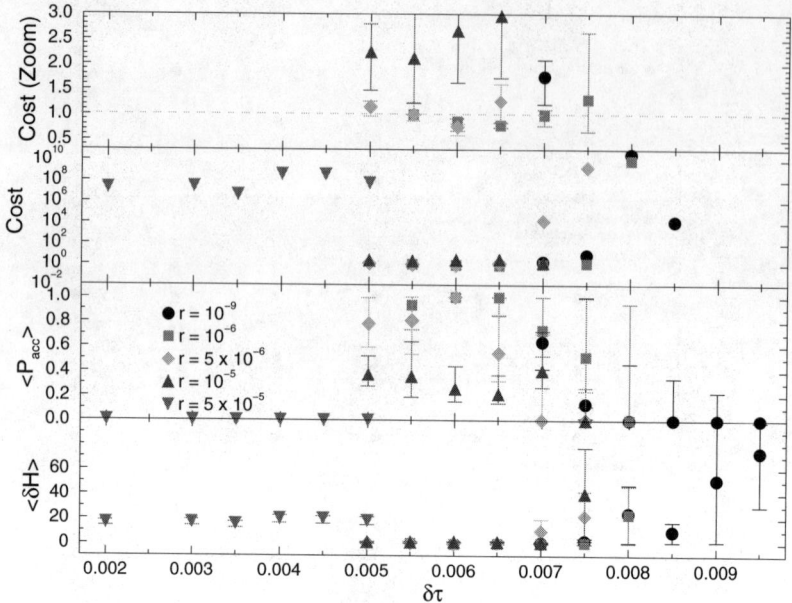

Fig. 4. The change in energy δH, acceptance probability $\langle P_{\text{acc}} \rangle$ and cost metric as a function of step size and solver residuum

one may be tempted to save some computational effort by solving the system with a less stringent convergence criteria.

In figure 4 we show the effects of relaxing the solver convergence criterion by plotting the energy change δH and the corresponding acceptance probability versus the step size $\delta\tau$ and for various values of the solver target residuum r. We also plot a cost function which combines the effects of changing he step–size and the resulting acceptance rates into a cost metric. The top graph is an enlargement of this cost function around its minimum. The cost is relative to the point with $\delta\tau = 0.0055$ and $r = 10^{-6}$. It can be seen that the window of opportunity (low cost) is not very great and within it, the minimum is very shallow.

We can see that as the target residuum r is increased to $r = 5 \times 10^{-4}$ we cannot achieve a value of δH which is of $O(1)$ for any step size, whereas even with a relatively stringent criterion of $r = 10^{-9}$ we cannot achieve a δH of $O(1)$ for step sizes greater than about $\delta\tau = 0.0085$.

5 Summary, Discussion and Conclusions

In this article we have discussed instabilities in the molecular dynamics integrators used in HMC simulations. We demonstrated the instability pedagogically in the simple harmonic oscillator system, for the simple leap-frog integrator and for the higher order integration schemes of [9, 10]. We showed that indeed the problem is worse for the higher order schemes as the instability is likely to become a problem there at smaller step sizes than for the leap-frog scheme.

We do not expect the integration schemes of [11] to alleviate the situation either, since they typically consist of making more coordinate update steps for each momentum update step to reduce the coefficient of the leading error term. The instabilities, however, arise from the combination of the force and the step-size in the momentum update step whose step–size is not decreased in typical applications of the scheme since the force computation makes this the more numerically expensive step.

We discussed the work of [4] and [5, 6] on instabilities in lattice QCD simulations. We adopted they hypothesis of [4] and showed data from [5, 6] that this picture was entirely consistent with what can be observed in large volume lattice QCD simulations. The key features of the picture are

- As the quark masses become lighter the fermionic forces increase
- The system goes unstable when the $F\delta\tau$ term reaches a critical value (in the momentum update step of the MD)
- As a result of the above, the energy change along an MD trajectory becomes very large, resulting in vanishing acceptance rates and therefore costly simulations.
- The problem can be cured by reducing the step-size $\delta\tau$.

We are particularly concerned with inexact, MD based algorithms such as the R and Φ algorithms used in staggered fermion simulations, since these algorithms do not contain an accept/reject step. Hence one may end up hitting the instability without knowing about it. The issue can be solved by performing a rigorous extrapolation of results in $\delta\tau$, which is not always carried out in practice.

References

1. Duane, S., Kennedy, A.D., Pendleton, B.J., Roweth, D.: Phys. Lett. **B195**, 216-222 (1987)
2. Peardon, M. J. These proceedings
3. K. Jansen and C. Liu, Nucl. Phys. B **453** (1995) 375, Liu, C., Jansen, K., Nucl. Phys. Proc. Suppl. **53**, 974–976 (1997), Liu, C., Jaster, A., Jansen, K., Nucl. Phys. **B524**, 603–617 (1998)
4. Edwards, R. G., Horváth, I., Kennedy, A. D., Nucl. Phys. Proc. Suppl. **53** 971–973 (1997), Edwards, R. G., Horváth, I., Kennedy, A. D., Nucl. Phys. **B484**, 375–402 (1997)
5. Joó, B.: Efficient Monte Carlo Simulation of Lattice QCD. PhD Thesis, University of Edinburgh, Edinburgh (2000)
6. Joó, B., Pendleton, B. J., Kennedy, A. D., Irving A. C., Sexton J. C., Pickles S. M., Booth, S. P., Phys. Rev. **D62**, 114501 (2000).
7. Metropolis, N. Rosenbluth A. W., Rosenbluth, M.N., Teller, A.H., Teller, E., J. Chem. Phys **21** 1087, (1953)
8. Kuti, J., Kennedy, A.D., Phys. Rev. Lett. **54**, 2473 (1985)
9. Campostrini, M., Rossi, P. Nucl. Phys **B329** 753 (1990)
10. Creutz, M., Gocksch, A. Phys. Rev. Lett **63** 9, (1989)
11. Sexton, J.C., Weingarten, D.H., Nucl. Phys. **B380** 665-677 (1992)
12. Allton C.R. *et. al.*, UKQCD Collaboration, Phys. Rev. **D65** 054502 (2002)

A Finite Baryon Density Algorithm

Keh-Fei Liu

Dept. of Physics and Astronomy, University of Kentucky, Lexington, KY 40506
USA liu@pa.uky.edu

Summary. I will review the progress toward a finite baryon density algorithm in the canonical ensemble approach which entails particle number projection from the fermion determinant. These include an efficient Padé-Z_2 stochastic estimator of the Tr log of the fermion matrix and a Noisy Monte Carlo update to accommodate unbiased estimate of the probability. Finally, I will propose a Hybrid Noisy Monte Carlo algorithm to reduce the large fluctuation in the estimated Tr log due to the gauge field which should improve the acceptance rate. Other application such as treating u and d as two separate flavors is discussed.

1 Introduction

Finite density is a subject of keen interest in a variety of topics, such as nuclear equation of state, neutron stars, quark gluon plasma and color superconductivity in nuclear physics and astrophysics, and high temperature superconductors in condensed matter physics. Despite recent advance with small chemical potential at finite temperature [1], the grand canonical approach with chemical potential remains a problem for finite density at zero temperature.

The difficulty with the finite chemical potential in lattice QCD stems from the infamous sign problem which impedes importance sampling with positive probability. The partition function for the grand canonical ensemble is represented by the Euclidean path-integral

$$Z_{GC}(\mu) = \int \mathcal{D}U \det M[U,\mu] e^{-S_g[U]},$$

where the fermion fields with fermion matrix M have been integrated to give the determinant. U is the gauge link variable and S_g is the gauge action. The chemical potential is introduced to the quark action with the $e^{\mu a}$ factor in the time-forward hopping term and $e^{-\mu a}$ in the time-backward hopping term. Here a is the lattice spacing. However, this causes the fermion action to be non-γ_5-Hermitian, i.e. $\gamma_5 M \gamma_5 \neq M^\dagger$. As a result, the fermion determinant det $M[U]$ is complex that leads to the sign problem.

2 Finite Chemical Potential

There are several approaches to avoid the sign problem. It was proposed by the Glasgow group [2] that the sign problem can be circumvented based on the expansion of the grand canonical partition function in powers of the fugacity variable $e^{\mu/T}$,

$$Z_{GC}(\mu/T, T, V) = \sum_{B=-3V}^{B=3V} e^{\mu/T\,B} Z_B(T, V),$$

where Z_B is the canonical partition function for the baryon sector with baryon number B. Z_{GC} is calculated with reweighting of the fermion determinant Since $Z_{GC}(\mu/T, T, V)$ is calculated with reweighting based on the gauge configuration with $\mu = 0$, it avoids the sign problem. However, this does not work, except perhaps for small μ near the finite temperature phase transition. We will dwell on this later in Sec. 3. This is caused by the 'overlap problem' [3] where the important samples of configurations in the $\mu = 0$ simulation has exponentially small overlap with those relevant for the finite density. To alleviate the overlap problem, a reweighting in multi-parameter space is proposed [4] and has been applied to study the end point in the T-μ phase diagram. In this case, the Monte Carlo simulation is carried out where the parameters in the set α_0 include $\mu = 0$ and β_c which corresponds to the phase transition at temperature T_c. The parameter set α in the reweighted measure include $\mu \neq 0$ and an adjusted β in the gauge action. The new β is determined from the Lee-Yang zeros so that one is following the transition line in the T-μ plane and the large change in the determinant ratio in the reweighting is compensated by the change in the gauge action to ensure reasonable overlap. This is shown to work to locate the transition line from $\mu = 0$ and $T = T_c$ down to the critical point on the 4^4 and $6^3 \times 4$ lattices with staggered fermions [4]. While the multi-parameter reweighting is successful near the transition line, it is not clear how to extend it beyond this region, particularly the $T = 0$ case where one wants to keep the β and quark mass fixed while changing the μ. One still expects to face the overlap problem in the latter case. It is shown [5] that Taylor expanding the observables and the reweighting factor leads to coefficients expressed in local operators and thus admits study of larger volumes, albeit still with small μ at finite temperature.

In the imaginary chemical potential approach, the fermion determinant is real and one can avoid the sign problem [6, 7, 8, 9, 10]. In practice, a reference imaginary chemical potential is used to carry out the Monte Carlo calculation and the determinants at other chemical potential values are calculated through a bosonic Monte Carlo calculation so that one can obtain the finite baryon partition function $Z_B(T, V)$ through the Fourier transform of the grand canonical partition function $Z_{GC}(\mu/T, T, V)$ [8]. However. this is problematic for large systems when the determinant cannot be directly calculated and it still suffers from the overlap problem. The QCD phase diagram has been studied with physical observables Taylor expanded and analytically continued to the real μ [9, 10]. Again, due to the overlap problem, one is limited to small real μ near the finite temperature phase transition.

3 Finite Baryon Density – a Canonical Ensemble Approach

An algorithm based on the canonical ensemble approach to overcome the overlap problem at zero temperature is proposed [11]. To avoid the overlap problem, one needs to lock in a definite nonzero baryon sector so that the exponentially large contamination from the zero-baryon sector is excluded. To see this, we first note that the fermion determinant is a superposition of multiple quark loops of all sizes and shapes. This can be easily seen from the property of the determinant

$$\det M = e^{\operatorname{Tr}\log M} = 1 + \sum_{n=1} \frac{(\operatorname{Tr}\log M)^n}{n!}.$$

Upon a hopping expansion of $\log M$, $\operatorname{Tr}\log M$ represents a sum of single loops with all sizes and shapes. The determinant is then the sum of all multiple loops. The fermion loops can be separated into two classes. One is those which do not go across the time boundary and represent virtual quark-antiquark pairs; the other includes those which wraps around the time boundary which represent external quarks and antiquarks. The configuration with a baryon number one which contains three quark loops wrapping around the time boundary will have an energy M_B higher than that with zero baryon number. Thus, it is weighted with the probability $e^{-M_B N_t a_t}$ compared with the one with no net baryons. We see from the above discussion that the fermion determinant contains a superposition of sectors of all baryon numbers, positive, negative and zero. At zero temperature where $M_B N_t a_t \gg 1$, the zero baryon sector dominates and all the other baryon sectors are exponentially suppressed. It is obvious that to avoid the overlap problem, one needs to select a definite nonzero baryon number sector and stay in it throughout the Markov chain of updating gauge configurations. To select a particular baryon sector from the determinant can be achieved by the following procedure [12]: first, assign an $U(1)$ phase factor $e^{-i\phi}$ to the links between the time slices t and $t+1$ so that the link $U/U^\dagger$ is multiplied by $e^{-i\phi}/e^{i\phi}$; then the particle number projection can be carried out through the Fourier transformation of the fermion determinant like in the BCS theory

$$P_N = \frac{1}{2\pi} \int_0^{2\pi} d\phi \, e^{-i\phi N} \det M(\phi) \tag{1}$$

where N is the net quark number, i.e. quark number minus antiquark number. Note that all the virtual quark loops which do not reach the time boundary will have a net phase factor of unity; only those with a net N quark loops across the time boundary will have a phase factor $e^{i\phi N}$ which can contribute to the integral in Eq. (1). Since QCD in the canonical formulation does not break $Z(3)$ symmetry, it is essential to take care that the ensemble is canonical with respect to triality. To this end, we shall consider the triality projection [12, 13] to the zero triality sector

$$\det{}_0 M = \frac{1}{3} \sum_{k=0,\pm 1} \det M(\phi + k 2\pi/3).$$

This amounts to limiting the quark number N to a multiple of 3. Thus the triality zero sector corresponds to baryon sectors with integral baryon numbers.

Another essential ingredient to circumvent the overlap problem is to stay in the chosen nonzero baryon sector so as to avoid mixing with the zero baryon sector

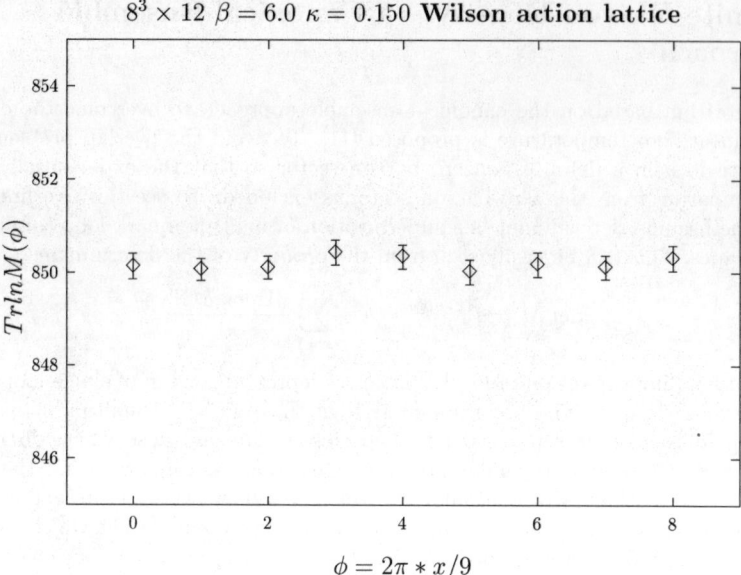

Fig. 1. $\mathrm{Tr}\log M(\phi)$ for a $8^3 \times 12$ quenched configuration with Wilson action as a function of ϕ.

with exponentially large weight. This can be achieved by performing the baryon number projection as described above *before* the accept/reject step in the Monte Carlo updating of the gauge configuration. If this is not done, the accepted gauge configuration will be biased toward the zero baryon sector and it is very difficult to project out the nonzero baryon sector afterwards. This is analogous to the situation in the nuclear many-body theory where it is known [14] that the variation after projection (Zeh-Rouhaninejad-Yoccoz method [15, 16]) is superior than the variation before projection (Peierls-Yoccoz method [17]). The former gives the correct nuclear mass in the case of translation and yields much improved wave functions in mildly deformed nuclei than the latter.

To illustrate the overlap problem, we plot in Fig.1 $\mathrm{Tr}\log M(\phi)$ for a configuration of the $8^3 \times 12$ quenched lattice with the Wilson action at $\beta = 6.0$ and $\kappa = 0.150$. This is obtained by the Padé-Z_2 estimator [19] of $\mathrm{Tr}\log M$ with 500 Z_2 noises where the entries of the noise vectors are chosen from the complex Z_2 group (i.e. 1 and -1) [20]. Z_2 noise is shown to produce the minimum variance and is thus optimal [23]. This will be explained in more detail later in Sec. 4.1. We see that the it is rather flat in ϕ indicating that the Fourier transform in Eq. (1) will mainly favor the zero baryon sector. On the other hand, at finite temperature, it is relatively easier for the quarks to be excited so that the zero baryon sector does not necessarily dominate other baryon sectors. Another way of seeing this is that the relative weighting factor $e^{-M_B N_t a_t}$ can be $O(1)$ at finite temperature. Thus, it should be easier to project out the nonzero baryon sector from the determinant. We plot in Fig. 2 a similarly obtained $\mathrm{Tr}\log M(\phi)$ for a configuration of the $8 \times 20^2 \times 4$ lattice with dynamical fermions at finite temperature with $\beta = 4.9$ and $\kappa = 0.182$. We see from the figure

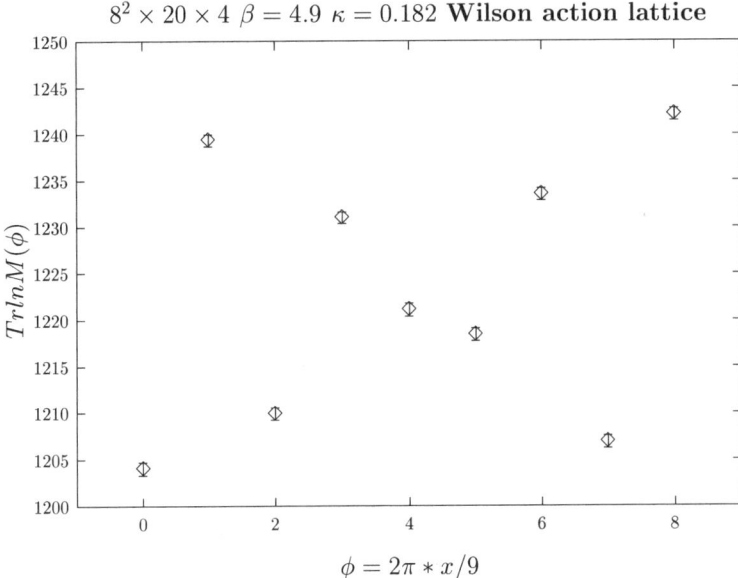

Fig. 2. Tr log $M(\phi)$ for a $8 \times 20^2 \times 4$ finite temperature configuration with dynamical fermions.

that there is quite a bit of wiggling in this case as compared to that in Fig. 1. This implies that it is easier to project out a nonzero baryon sector through the Fourier transform at finite temperature.

4 Noisy Monte Carlo with Fermion Determinant

In order to implement the canonical ensemble approach, it is clear that one needs to evaluate the fermion determinant for the purpose of particle projection. Since the pseudofermion approach does not give the determinant in the Markov process, it is not applicable. In view of the fact that it is impractical to calculate the determinant directly for realistic volumes, a Monte Carlo algorithm which accommodates an unbiased estimate of the probability and an efficient way to estimate the determinant are necessary for the finite baryon density calculation.

A Noisy Monte Carlo algorithm [18] with Padé-Z_2 estimates [19, 20] of the Tr log of the fermion matrix are developed toward this goal and a numerical simulation with Wilson dynamical fermion is carried out [21]. We shall summarize the progress made so far.

The QCD partition function can be written in the form

$$Z = \int dU\, e^{-S_g(U)} \int \prod_{i=1}^{\infty} d\eta_i\, P^\eta(\eta_i)$$

$$\times \prod_{k=2}^{\infty} d\rho_k\, P^\rho(\rho_k)\, f(U, \eta, \rho),$$

where $S_g(U)$ is the gauge action. $f(U, \eta, \rho)$ stands for $f(U, \{\eta_i\}, \{\rho_k\})$ which is an unbiased stochastic estimator [22] of the fermion determinant $e^{\mathrm{Tr}\log M}$ via an infinite number of auxiliary variables ρ_k and η_i. $P^\eta(\eta_i) = \delta(|\eta_i| - 1)$ is the distribution for the Z_2 noise η_i and $P^\rho(\rho_k) = \theta(\rho_k) - \theta(\rho_k - 1)$ is the flat distribution for $0 \leq \rho_k \leq 1$. With $f(U, \{\eta_i\}, \{\rho_k\})$ being the stochastic expansion

$$f(U, \{\eta_i\}, \{\rho_k\}) = 1 + \left\{ x_1 + \theta(1 - \rho_2) \left\{ x_2 + \theta\left(\frac{1}{3} - \rho_3\right) \{x_3 + \ldots \right.\right.$$

$$\left.\left. \ldots + \theta\left(\frac{1}{n} - \rho_n\right)\{x_n + \ldots\}\right\}\right\}\right\} \quad (2)$$

where $x_i = \eta_i^\dagger \ln M(U) \eta_i$, one can verify [22] that

$$\prod_{i=1}^{\infty} d\eta_i P^\eta(\eta_i) \prod_{k=2}^{\infty} d\rho_k P^\rho(\rho_k) \langle f(U, \{\eta_i\}, \{\rho_k\}) \rangle = e^{\mathrm{Tr}\ln M(U)},$$

and the stochastic series terminates after e terms on the average.

Since the estimator $f(U, \eta, \rho)$ can be negative due to the stochastic estimation, the standard treatment is to absorb the sign into the observables, i.e.

$$\langle \mathcal{O} \rangle_P = \frac{\langle \mathcal{O}\, \mathrm{sgn}(P) \rangle_{|P|}}{\langle \mathrm{sgn}(P) \rangle_{|P|}}.$$

With the probability for the gauge link variable U and noise $\xi \equiv (\eta, \rho)$ written as $P(U, \xi) \propto P_1(U) P_2(U, \xi) P_3(\xi)$ with

$$P_1(U) \propto e^{-S_g(U)}$$

$$P_2(U, \xi) \propto |f(U, \xi)|$$

$$P_3(\xi) \propto \prod_{i=1}^{\infty} P^\eta(\eta_i) \prod_{k=2}^{\infty} P^\rho(\rho_k),$$

the following two steps are needed to prove detailed balance [18, 21].

(a) Let $T_1(U, U')$ be the ergodic Markov matrix satisfying detailed balance with respect to P_1, in other words $P_1(U) T_1(U, U') dU = P_1(U') T_1(U', U) dU'$. Then the transition matrix

$$T_{12}(U, U') = T_1(U, U') \min\left[1, \frac{P_2(U', \xi)}{P_2(U, \xi)}\right]$$

satisfies detailed balance with respect to the $P_1(U) P_2(U, \xi)$ (with ξ fixed).

(b) The transition matrix

$$T_{23}(\xi, \xi') = P_3(\xi') \min\left[1, \frac{P_2(U, \xi')}{P_2(U, \xi)}\right]$$

satisfies detailed balance with respect to $P_2(U,\xi)P_3(\xi)$ (with U fixed).

From (a), (b) it follows that T_{12} and T_{23} keep the original distribution $P(U,\xi)$ invariant and interleaving them will lead to an ergodic Markov process with the desired fixed point.

4.1 Padé - Z_2 Estimator of $\text{Tr} \ln M$ with Unbiased Subtraction

In Eq. (2), one needs to calculate $x_i = \eta_i^\dagger \ln M(U) \eta_i$ in the stochastic series expansion of the fermion determinant. An efficient method is developed to calculate it [19]. First of all, the logarithm is approximated using a Padé approximation, which after the partial fraction expansion, has the form

$$\ln M(U,\kappa) \approx R_M(U) \equiv b_0 \; I + \sum_{i=1}^{N_P} b_i \left(M(U,\kappa) + c_i \; I\right)^{-1} \qquad (3)$$

where N_P is the order of the Padé approximation, and the constants b_i and c_i are the Padé coefficients. In our implementation we have used an 11-th order approximation whose coefficients are tabulated in [19]. The traces of $\ln M$ are then estimated by evaluating bilinears of the form $\eta^\dagger R_M(U) \eta$. If the components of η are chosen from the complex Z_2 group, then the contributions to the variance of these bilinears come only from off diagonal elements of $R_M(U)$ [23, 20]. In this sense, Z_2 noise is optimal and has been applied to the calculation of nucleon matrix elements involving quark loops [24]. An effective method reducing the variance is to subtract off a linear combination of traceless operators from $R_M(U)$ and to consider

$$E[\text{Tr } R_M(U), \eta] = \eta^\dagger \left(R_M(U) - \alpha_i \mathcal{O}_i\right) \eta \; .$$

Here the $\mathcal{O}_i$ are operators with $\text{Tr } \mathcal{O}_i = 0$. Clearly since the $\mathcal{O}_i$ are traceless they do not bias the estimators. The α_i are constants that can be tuned to minimize the fluctuations in $E[\text{Tr } R_M(U), \eta]$.

With other types of noise such as Gaussian noise, the variance receives contributions from diagonal terms which one cannot subtract off. In this case, the unbiased subtraction scheme described here is ineffective. In practice, the $\mathcal{O}_i$ are constructed by taking traceless terms from the hopping parameter expansion for $M^{-1}(U)$. It is shown for Wilson fermions on a $8^3 \times 12$ lattice at $\beta = 5.6$, these subtractions can reduce the noise coming from the terms $(M(U) + c_i)^{-1}$ in Eq. (3) by a factor as large as 37 for $\kappa = 0.150$ with 50 Z_2 noises [19].

4.2 Implementation of the Noisy Monte Carlo Algorithm

The Noisy Monte Carlo algorithm has been implemented for the Wilson dynamical fermion with pure gauge update (Kentucky Noisy Monte Carlo Algorithm) for an 8^4 lattice with $\beta = 5.5$ and $\kappa = 0.155$ [21]. Several tricks are employed to reduce the fluctuations of the $\text{Tr} \ln M$ estimate and increase the acceptance. These include shifting the $\text{Tr} \ln M$ with a constant, $\Delta\beta$ shift [25], and splitting the $\text{Tr} \ln M$ with N 'fractional flavors'. After all these efforts, the results are shown to agree with those from the HMC simulation. However, the autocorrelation is very long and the acceptance rate is low. This has to do with the fact that $\text{Tr} \ln M$ is an extensive quantity which is proportional to volume and the stochastic series expansion of e^x

converges for $x \leq 6$ for a sample with the size of $\sim 10^3 - 10^4$. This is a stringent requirement which requires the fractional flavor number $N \geq 15$ for this lattice. This can be seen from the distribution of $x = \sum_f (\text{Tr} R_M^f(U) - \lambda^f Plaq - x_0^f)/N$ in Fig. 3 which shows that taking N to be 15, 20, and 25, the largest x value is less than 6.

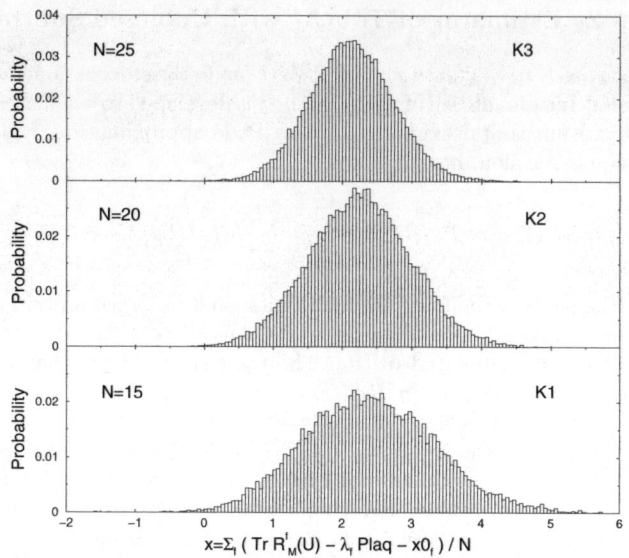

Fig. 3. Distributions of x for the three noisy simulations

As the volume increases, this fractional flavor needs to be larger to keep x smaller than 6 for a sample of the size $\sim 10^3 - 10^4$. At the present volume (8^4), the autocorrelation is already much longer than that of HMC, it is going to be even less efficient for larger volumes. This is generic for the Noisy Monte Carlo algorithm which scales with volume as V^2, while HMC scales as $V^{5/4}$.

5 Hybrid Noisy Monte Carlo Algorithm – a New Proposal

It is clear that the inefficiency of the Noisy Monte Carlo algorithm for the fermion determinant is due to the large fluctuation of the $\text{Tr}\ln M$ estimator from one gauge configuration to the next. We shall propose a combined Hybrid Monte Carlo (HMC) and Noisy Monte Carlo (NMC) to remove such fluctuations in the context of the finite density.

With the baryon number projection discussed in Sec. 3, we can write the partition function for the finite baryon sector with B baryons as

$$Z_B = \int dp\, dU\, d\phi^\dagger\, d\phi\, e^{-p^2/2 - S_g(U) + \phi^\dagger (M)^{-1}\phi} \left\{ \frac{\frac{1}{2\pi} \int_0^{2\pi} d\theta\, e^{-i3B\theta} \det M(\theta)}{\det M(\theta = 0)} \right\}.$$

In this case, one can update the momentum p, the gauge link variable U and the pseudofermion field ϕ via HMC and then interleave with NMC for updating the determinant ratio

$$R = \frac{\frac{1}{2\pi}\int_0^{2\pi} d\theta\, e^{-i3B\theta} \det M(\theta)}{\det M(\theta=0)}. \tag{4}$$

As described in Sec. 4, NMC involves two Metropolis accept/reject steps to update the ratio with the Padé - Z_2 estimator of the Tr ln difference of the determinants, i.e. $\text{Tr}(\ln M(\theta) - \ln M(\theta=0))$. It is pointed out [26] that for zero temperature, one can approximate the continuous integral over θ with a discrete sum incorporating triality zero projection [12, 13] so that the partition function is a mixture of different Z_B for different baryon number B. In other words, the approximation

$$\frac{1}{2\pi}\int_0^{2\pi} d\theta\, e^{-i3B\theta} \det M(\theta) \longrightarrow \frac{1}{3B_N} \sum_{k=0}^{3B_N-1} e^{-i\frac{2\pi kB}{3B_N}} \det M(2\pi k B/3B_N), \tag{5}$$

leads to the mixing of the baryon sector B with those of $B \pm B_N, B \pm 2B_N$.... If B is small and $B_N > B$, then the partition will be dominated by Z_B with small mixture from $Z_{B\pm B_N}, Z_{B\pm 2B_N}, ...$ For example, if we take $B = 1$ and $B_N = 5$, the discrete approximation gives an admixture of partition function with baryon number $B = 1, 5, 11, -4, -9, ...$ At zero temperature, the partition function Z_B behaves like $e^{-Bm_N N_t a_t}$, one expects that the mixing due to baryons other than $B = 1$ will be exponentially suppressed when $m_n N_t a_t > 1$.

Two points need to be stressed. First of all, it is crucial to project out the particle number in Eq. (5) before the Metropolis accept/reject step in order to overcome the overlap problem. Secondly, given that the ratio R in Eq. (4) is replaced with a discrete sum

$$\overline{R} = \frac{1}{3B_N} \sum_{k=0}^{3B_N-1} e^{-i\frac{2\pi kB}{3B_N}} e^{\text{Tr}(\ln M(2\pi kB/3B_N) - \ln M(0))},$$

which involves the difference between the $\text{Tr}\ln M(2\pi kB/3B_N)$ and $\text{Tr}\ln M(0)$, it takes out the fluctuation due to the gauge configuration which plagued the Kentucky Noisy Monte Carlo simulation in Sec. 4. Furthermore, the Tr ln difference is expected to be $O(1)$ as seen from Fig. 1. If that is indeed the case, it should lead to a better convergence of the stochastic series expansion in Eq. (2) and the algorithm scales with volume the same as HMC.

5.1 Another Application

Here we consider another possible application of the Hybrid Noisy Monte Carlo algorithm. HMC usually deals with two degenerate flavors. However, nature comes with 3 distinct light flavors – u, d and s. To consider u and d as separate flavors, one can perform HMC with two degenerate flavors at the d quark mass and then employ NMC to update the determinant ratio

$$R_{ud} = \frac{\det M_d^\dagger \det M_u}{\det M_d^\dagger \det M_d} = e^{\text{Tr}(\ln M_u - \ln M_d)}. \tag{6}$$

Since both the u and d masses are much smaller than Λ_{QCD}, $\text{Tr}(\ln M_u - \ln M_d)$ should be small. If the Tr ln difference is small enough (e.g. $O(1)$) so that the acceptance rate is high, it could be a feasible algorithm for treating u and d as distinct flavors so that realistic comparison with experiments can be done someday. It is shown recently that the Rational Hybrid Monte Carlo Algorithm (RHMC) [27, 28] works efficiently for two flavor staggered fermions. It can be applied to single flavors for Wilson, domain wall, or overlap fermions at the cost of one pesudofermion for each flavor. We should point out that, in comparison, the Hybrid Noisy approach discussed here saves one pseudofermion, but at the cost of having to update the determinant ratio R_{ud} in Eq. (6).

While we think that the Hybrid Noisy Monte Carlo algorithm proposed here might overcome the overlap and the low acceptance problems and the determinant $\det M(\theta)$ is real in this approach, the fact that the Fourier transform in Eq. (5) involves the baryon number B may still lead to a sign problem in the thermodynamic limit when B and V are large and also when the temperature T is low. However, as an initial attempt, we are more interested in finding out if the algorithm works for a small B such as 1 or 2 in a relatively small box and near the deconfinement phase transition temperature. We should point out that in a special case of two degenerate flavors with the same baryon number, the ratio R in Eq. (4) becomes

$$R = \frac{(\frac{1}{2\pi}\int_0^{2\pi} d\theta\, e^{-i3B\theta} \det M(\theta))^2}{\det M^\dagger M(\theta = 0)},$$

which is positive and is thus free of the sign problem.

6 Acknowledgment

This work is partially supported by DOE grants DE-FG05-84ER0154 and DE-FG02-02ER45967. The author would like to thank the organizers of the "Third International Workshop on QCD and Numerical Analysis' for the invitation to attend this stimulating workshop. He also thanks B. Joó, S.J.Dong, I. Horváth, and M. Faber for stimulating discussions of the subject.

References

1. For a review, see for example, S. D. Katz, hep-lat/0310051.
2. I. M. Barbour, S. E. Morrison, E. G. Klepfish, J. B. Kogut, and M.-P. Lombardo, *Nucl. Phys. (Proc. Suppl.)* **60A**, 220 (1998); I.M. Barbour, C.T.H. Davies, and Z. Sabeur, *Phys. Lett.* **B215**, 567 (1988).
3. M. Alford, *Nucl. Phys. B (Proc. Suppl.)* **73**, 161 (1999).
4. Z. Fodor and S.D. Katz, *Phys. Lett.***B534**, 87 (2002); *JHEP* **0203**, 014 (2002).
5. C. R. Allton et al., *Phys. Rev.* **D66**, 074507 (2002).
6. E. Dagotto, A. Moreo, R. Sugar, and D. Toussaint, *Phys. Rev.* **B41**, 811 (1990).
7. N. Weiss, *Phys. Rev.* **D35**, 2495 (1987); A. Hasenfratz and D. Toussaint, *Nucl. Phys.* **B371**, 539 (1992).
8. M. Alford, A. Kapustin, and F. Wilczek, *Phys. Rev.* **D59**, 054502 (2000).

9. P. deForcrand and O. Philipsen, *Nucl. Phys.* **B642**, 290 (2002); ibid, **B673**, 170 (2003).
10. M. D'Elia and M.-P. Lombardo, *Phys. Rev.* **D67**, 014505 (2003).
11. K.F. Liu, *Int. Jour. Mod. Phys.* **B16**, 2017 (2002).
12. M. Faber, O. Borisenko, S. Mashkevich, and G. Zinovjev, *Nucl. Phys.* **B444**, 563 (1995).
13. M. Faber, O. Borisenko, S. Mashkevich, and G. Zinovjev, *Nucl. Phys. B(Proc. Suppl.)* **42**, 484 (1995).
14. For reviews on the subject see for example C.W. Wong, *Phys. Rep.* **15C**, 285 (1975); D.J. Rowe, Nuclear Collective Motion, Methuen (1970); P. Ring and P. Schuck, The Nuclear Many-Body Problem, Springer-Verlag (1980).
15. H.D. Zeh, *Z. Phys.* **188**, 361 (1965).
16. H. Rouhaninejad and J. Yoccoz, *Nucl. Phys.* **78**, 353 (1966).
17. R.E. Peierls and J. Yoccoz, *Proc. Phys. Soc. (London)* **A70**, 381 (1957).
18. L. Lin, K. F. Liu, and J. Sloan, *Phys. Rev.* **D61**, 074505 (2000), [hep-lat/9905033]
19. C. Thron, S. J. Dong, K. F. Liu, H. P. Ying, *Phys. Rev.* **D57**, 1642 (1998); K.F. Liu, *Chin. J. Phys.* **38**, 605 (2000).
20. S. J. Dong and K. F. Liu, *Phys. Lett.* **B 328**, 130 (1994).
21. B. Joó, I. Horváth, and K.F. Liu, *Phys. Rev.* **D67**, 074505 (2003).
22. G. Bhanot, A. D. Kennedy, *Phys. Lett.* **157B**, 70 (1985).
23. S. Bernardson, P. McCarty and C. Thron, *Comp. Phys. Commun.* **78**, 256 (1994).
24. S.J. Dong, J.-F. Lagaë, and K.F. Liu, *Phys. Rev. Lett.* **75**, 2096 (1995); S.J. Dong, J.-F. Lagaë, and K.F. Liu, *Phys. Rev.* **D54**, 5496 (1996); S.J. Dong, K.F. Liu, and A. G. Williams, *Phys. Rev.* **D58**, 074504 (1998).
25. J.C. Sexton and D.H. Weingarten, *Nucl. Phys.* **B(Proc. Suppl.)42**, 361 (1995).
26. M. Faber, private communication.
27. I. Horváth, A. Kennedy, and S. Sint, *Nucl. Phys.* **B(Proc. Suppl.)** **73**, 834 (1999).
28. M. Clark and A. Kennedy, hep-lat/0309084.

The Nucleon Mass in Chiral Effective Field Theory

Ross D. Young, Derek B. Leinweber, and Anthony W. Thomas

Special Research Centre for the Subatomic Structure of Matter,
and Department of Physics, University of Adelaide,
Adelaide SA 5005, Australia

Summary. We review recent analysis that has developed finite-range regularised chiral effective field theory as an efficient tool for studying the quark mass variation of QCD observables. Considering a range of regularisation schemes, we study both the expansion of the nucleon mass about the chiral limit and the practical application of extrapolation for modern lattice QCD.

1 Introduction

State-of-the-art lattice calculations in dynamical QCD are typically restricted to the simulation of light quark masses which are at least an order of magnitude larger than their physical values. This necessitates an extrapolation in quark mass in order to make comparison between theory and experiment. This extrapolation to the chiral regime is non-trivial due to non-analytic variation of hadron properties with quark mass.

It has recently been established that chiral extrapolation can be accurately performed with the use of finite-range regularisation (FRR) in chiral effective field theory [1, 2]. We highlight some of the features of finite-range regularisation and demonstrate the application to the extrapolation of the nucleon mass.

2 Effective Field Theory and Renormalisation

Chiral perturbation theory (χPT) is a low-energy effective field theory (EFT) of QCD, for a review see Ref. [3]. Using this effective field theory, the low-energy properties of hadrons can be expanded about the limit of vanishing momenta and quark mass. In particular, in the context of the extrapolation of lattice data, χPT provides an expansion in m_q about the chiral limit. Equivalently, this can be translated to an expansion in the scale-independent pion mass, m_π, through the Gell-Mann–Oakes–Renner (GOR) relation $m_\pi^2 \propto m_q$

[4]. Goldstone boson loops play an especially important role in the theory as they give rise to non-analytic behaviour as a function of quark mass. The low-order, non-analytic contributions arise from the pole in the Goldstone boson propagator and hence are *model-independent* [5]. It is the renormalisation of these non-analytic contributions which form the primary issue of discussion in this paper.

Renormalisation in effective field theories has been discussed in the pedagogic introduction given a few years ago by Lepage [6]. The problem considered in this lecture was to develop an effective field theory for the solution of the Schrödinger equation for an unknown potential. The key physical idea of the effective field theory is to introduce an energy scale, λ, above which one does not attempt to understand the physics. Considering one does not pretend to control physics above the scale λ, one should not include momenta above λ in computing radiative corrections. Instead, one introduces renormalization constants which depend on (or "run with") the choice of cut-off λ so that, to the order one works, physical results are independent of λ.

In this study Lepage reconstructs the potential based on the low-energy EFT, using both dimensional regularisation and a cutoff scheme, which we refer to as FRR. It is found that, to the working order of the low-energy expansion, the FRR scheme has vastly improved properties over the dimensionally regulated counterpart. The poor performance of dimensional regularisation can be understood in terms of the incorrect contributions associated with short-distance physics. In particular, dimensional regularization involves integrating loop momenta in the effective field theory over momenta all the way to infinity – way beyond the scale where the effective theory has any physical significance.

In the context of fitting lattice data to the chiral expansion of the EFT we are led to the evaluation of loop corrections using a finite scale in the regularisation prescription. This scale is chosen to lie below that where the low-energy theory breaks down. Any possible residual dependence on the specific choice of mass scale (and form of regulator function) will be eliminated by fitting the renormalisation constants to nonperturbative QCD – in this case data obtained from lattice QCD simulations. The quantitative success of applying the method is to be judged by the extent to which physical results extracted from the analysis are indeed independent of the regulator.

3 Chiral Expansion of the Nucleon Mass

We now turn to an investigation of the effective rate of convergence of the chiral expansions obtained using different functional forms for the regulator. We also make a comparison with the results obtained using a dimensionally regulated approach.

Lattice QCD provides us with a reliable, non-perturbative method for studying the variation of M_N with m_q. To constrain the free parameters of the expansion we take as input both the physical nucleon mass and recent lattice QCD results of the CP-PACS Collaboration [7] and the JLQCD Collaboration [8]. For details of the expansion and the parameters see Ref. [1].

The physical scale of the lattice simulations is set using Sommer scale, $r_0 = 0.5\,\text{fm}$ [9, 10]. For the purpose of chiral extrapolation it is important that the method for setting the scale does not contaminate the chiral physics being investigated. Being based on the static quark potential, which is rather insensitive to chiral physics, the Sommer scale is ideal in the present context.

We describe the expansion of the nucleon mass at next-to-leading non-analytic order by

$$M_N = a_0 + a_2 m_\pi^2 + a_4 m_\pi^4 + a_6 m_\pi^6 + \chi_\pi I_\pi(m_\pi, 0) + \chi_{\pi\Delta} I_\pi(m_\pi, \Delta), \quad (1)$$

where the terms involving χ_π and $\chi_{\pi\Delta}$ describe the one loop chiral corrections to the nucleon mass for $N \to N\pi$ and $N \to \Delta\pi$, respectively. The chiral couplings are fixed to their phenomenological values

$$\chi_\pi = -\frac{3}{32\pi f_\pi^2} g_A^2, \qquad \chi_{\pi\Delta} = -\frac{3}{32\pi f_\pi^2} \frac{32}{25} g_A^2.$$

The corresponding loop integral, in the heavy-baryon limit, is given by

$$I_\pi(m_\pi, \Delta) = \frac{2}{\pi} \int_0^\infty dk \, \frac{k^4}{\sqrt{k^2 + m_\pi^2}[\Delta + \sqrt{k^2 + m_\pi^2}]},$$

with the nucleon–delta mass splitting defined at the physical value, $\Delta = M_\Delta - M_N$.

The form of Eq. (1) is unrenormalised and divergent behaviour of the loop integrals must be regularised. We study a variety of regularisation schemes for the chiral expansion. We allow each scheme to serve as a constraint curve for the other methods. In particular, we generate six different constraint curves, for each regularisation, that describe the quark mass dependence of M_N. The first case corresponds to the truncated power series obtained through the dimensionally regulated (DR) approach. The second procedure takes a similar form but we maintain the complete branch point (BP) structure at $m_\pi = \Delta$ [11]. Finally, we use four different functional forms for the finite-ranged, ultraviolet vertex regulators — namely the sharp-cutoff (SC), monopole (MON), dipole (DIP) and Gaussian (GAU). We refer to Ref. [1] for details.

As highlighted above, provided one regulates the effective field theory below the point where new short distance physics becomes important, the essential results will not depend on the cut-off scale [6]. We use knowledge learned in Ref. [1] as a guide for the appropriate scales for each of the regulator forms. In particular, we choose regulator masses of 0.4, 0.5, 0.8 and 0.6 GeV for the sharp, monopole, dipole and Gaussian regulators, respectively.

In each case, both DR-based and FRR, we fit four free parameters, $a_{0,2,4,6}$ to constrain the expansion of the nucleon mass using the physical nucleon mass

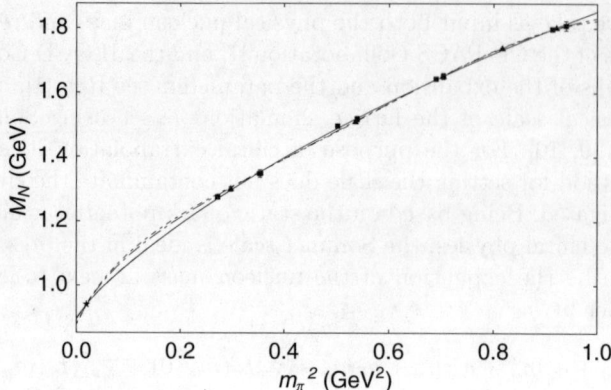

Fig. 1. Various regularisation schemes providing constraint curves for the variation of M_N with pion mass. The short dash curve corresponds to DR and the long dash curve to the BP approach, as discussed in the text. The four FRR curves (solid) are indistinguishable at this scale.

and the lattice data discussed above. The values of the non-analytic contributions are fixed to their model-independent values. The resultant curves are displayed in Fig. 1. All of the curves are in good agreement with each other and are able to give an accurate description of the lattice data and match the physical value of M_N.

The best fit parameters for our constraint curves, $M(m_\pi)$, are shown in Table 1. It is important to note that the parameters listed in this table are

Regulator	a_0	a_2	a_4	a_6
DR	0.877	4.06	5.57	-3.24
BP	0.821	4.57	8.92	-2.02
SC	1.02	1.15	-0.359	0.055
MON	1.58	0.850	-0.162	-0.007
DIP	1.22	0.925	-0.177	-0.009
GAU	1.11	1.03	-0.265	0.024

Table 1. Coefficients of the residual series expansion of the nucleon mass for various regularization schemes, as explained in the text. All units in appropriate powers of GeV.

bare quantities and hence *renormalisation scheme dependent*. To make rigorous comparison to the parameters of the effective field theory, the loop contributions must be Taylor expanded about $m_\pi = 0$ in order to yield the renormalisation for each of the coefficients in the quark mass expansion about the chiral limit [1, 12]. In any regularisation scheme, the *renormalised* chiral expansion is given by

$$M_N = c_0 + c_2 m_\pi^2 + \chi_\pi m_\pi^3 + c_4 m_\pi^4 - \chi_{\pi\Delta} \frac{3}{4\pi\Delta} m_\pi^4 \log \frac{m_\pi}{\mu} + \dots.$$

We note that the definition of the parameter c_4 is implicitly dependent on the scale of the logarithm, which we set to $\mu = 1\,\text{GeV}$.

The renormalised expansion coefficients for each of the regularisation schemes are shown in Table 2. The most remarkable feature of Table 2 is the very close agreement between the values of the renormalised coefficients, especially for the finite-range regularisation schemes. For example, whereas the variation in a_0 between all four FRR schemes is 50%, the variation in c_0 is a fraction of a percent. For a_2 the corresponding figure is 30% compared with less than 9% variation in c_2. If one excludes the less physical sharp cut-off (SC) regulator, the monopole, dipole and Gaussian results for c_2 vary by only 2%. Finally, for c_4 the agreement is also good for the latter three schemes.

It is the higher order terms that highlight why the FRR schemes are so efficient. Considering a_4 and c_4 we observe that the renormalised coefficients are consistently very large for the three smooth FRR schemes, whereas the bare coefficients of the residual expansion are two orders of magnitude smaller! That is, once the non-analytic corrections are evaluated with a FRR the residual series (ie. the a_i coefficients) shows remarkably improved convergence and hence the truncation is shown to be reliable.

The dimensional regularisation schemes do not offer any sign of convergence, where the bare expansion coefficients are much larger — a factor of 30 in the case of a_4. It is interesting to observe, that although these coefficients are large they are not large enough to reproduce consistent values with the smooth FRR results reported in Table 2.

Regulator	c_0	c_2	c_4
DR	0.877	4.10	5.57
BP	0.881	3.84	7.70
SC	0.895	3.02	14.0
MON	0.898	2.78	23.5
DIP	0.897	2.83	21.7
GAU	0.897	2.83	21.2

Table 2. The renormalised chiral expansion parameters for different regularisation schemes. All units in appropriate powers of GeV.

We take each of our constraint curves and then do a one-to-one comparison to determine the efficiency of any alternative scheme in describing the constraint. Being based on the same EFT, all of the regularisation prescriptions have precisely the same structure in the limit $m_\pi \to 0$. As our test window moves out to larger pion mass the accurate reproduction of the low-energy constants serves as a test of the efficiency of the regularisation.

Although the analysis is modified slightly, we find results which are equivalent to the study in Ref. [1]. All regularisation schemes are demonstrated to be equivalent for pion masses below $m_\pi^2 \sim 0.4\,\mathrm{GeV}^2$. Above this scale, accurate reproduction of the low-energy constants is restricted between the different regulators. It is evident that the DR-type schemes and the FRR schemes cannot give consistent results over such large ranges of pion mass. The agreement between the FRR schemes alone is excellent over the entire range considered. We also note that our choice of scale, Λ, was set to the best phenomenological value given the guidance of Ref. [1]. It is worth noting that if one tuned Λ it could exactly reproduce the BP curves, as the BP dimensional regularisation is identical to FRR with $\Lambda \to \infty$.

It is evident that if one wishes to study the chiral expansion based on DR chiral EFT in lattice QCD one must only work with data below $m_\pi^2 \sim 0.4\,\mathrm{GeV}^2$. In fact, one would need a large number of accurate points in this region in order to reliably fix four fitting parameters. In addition, data very near the physical pion mass would be necessary. On the other hand, it is apparent that by using the improved convergence properties of the FRR schemes one can use lattice data in the region up to $1.0\,\mathrm{GeV}^2$. This is a regime where we already have impressive data from CP-PACS [7], JLQCD [8], MILC [13] and UKQCD [14]. In particular, the FRR approach offers the ability to extract reliable fits where the low-mass region is excluded from the available data. The consideration of this practical application to the extrapolation of lattice data will be discussed further below.

4 Chiral Extrapolation

In the previous discussion all we were interested in was the expansion about the chiral limit and hence we obtained the best phenomenological curves by including the physical nucleon mass as a constraint. Here we examine the *extrapolation* of real lattice QCD results. Thus the extrapolation gives physical predictions of nonperturbative QCD from lattice simulations.

We examine the same two sets of lattice data, as discussed above. For the four FRR considered, we investigate the residual regulator dependence on the chiral extrapolation. For comparison, we also show extrapolations based on the DR-based EFT. As in the previous section, we allow the coefficients of the analytic terms up to m_π^6 to be determined by the lattice data. The regulator masses are once again fixed to their preferred values. We show the best fits in Fig. 2.

All finite-range regularisation prescriptions give almost identical results, with the discrepancies beyond the resolution of Fig. 2. The extrapolation of lattice QCD data based on FRR χPT establishes that that the extraction of physical properties can be systematically carried out using simulation results at quark masses significantly larger than nature.

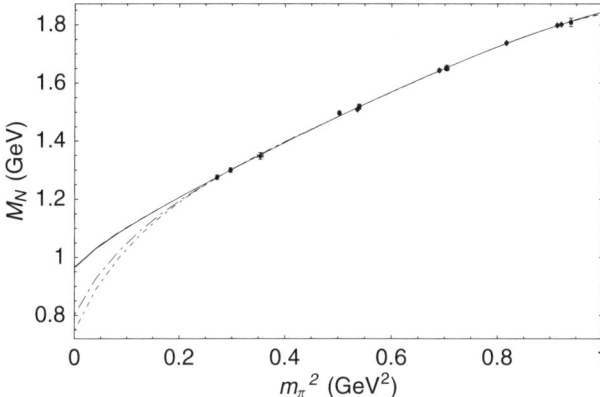

Fig. 2. Extrapolation of lattice data for various regularisation schemes. The four *indistinguishable*, solid curves show the FRR extrapolations. The short dash-dot curve corresponds to the DR scheme and the long dash-dot that of BP.

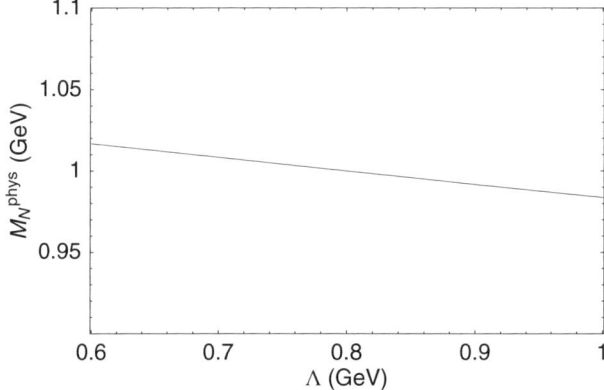

Fig. 3. The extrapolated nucleon mass for varying dipole mass, Λ.

The insensitivity of the extrapolation for a wide range of regulator scale is shown in Fig. 3. We show the variation of the extrapolated nucleon mass, at the physical pion mass, for dipole masses ranging between 0.6 to 1.0 GeV. It is observed that the residual uncertainty introduced by the regulator scale is less than 2%. Compared to the 13% statistical error in the extrapolation, this discrepancy induced by the choice of scale is insignificant.

5 Conclusion

We have demonstrated the accuracy and reliability of extrapolations of lattice QCD based on FRR chiral EFT. The extrapolation distance is still very large, meaning that the statistical uncertainties do not yet provide accurate predictive power. As dynamical simulations of lattice QCD with improved actions approach lighter quark masses, FRR χPT will provide for accurate extraction of physical QCD properties.

This work was supported by the Australian Research Council.

References

1. R. D. Young *et al.*, Prog. Part. Nucl. Phys. **50**, 399 (2003) [arXiv:hep-lat/0212031].
2. D. B. Leinweber, A. W. Thomas and R. D. Young, Phys. Rev. Lett. **92**, 242002 (2004) [arXiv:hep-lat/0302020].
3. V. Bernard, N. Kaiser and U.-G. Meissner, Int. J. Mod. Phys. **E4**, 193 (1995).
4. M. Gell-Mann, R.J. Oakes and B. Renner, Phys. Rev. 175 (1968) 2195.
5. L. F. Li and H. Pagels, Phys. Rev. Lett. **26**, 1204 (1971).
6. G. P. Lepage, arXiv:nucl-th/9706029.
7. A. Ali Khan *et al.* [CP-PACS Collaboration], Phys. Rev. D **65**, 054505 (2002) [Erratum-ibid. D **67**, 059901 (2003)] [arXiv:hep-lat/0105015].
8. S. Aoki *et al.* [JLQCD Collaboration], Phys. Rev. D **68**, 054502 (2003) [arXiv:hep-lat/0212039].
9. R. Sommer, Nucl. Phys. B411 (1994) 839, hep-lat/9310022.
10. R.G. Edwards *et al.*, Nucl. Phys. B517 (1998) 377, hep-lat/9711003.
11. M. K. Banerjee and J. Milana, Phys. Rev. D **54**, 5804 (1996) [arXiv:hep-ph/9508340].
12. J. F. Donoghue *et al.*, Phys. Rev. D **59**, 036002 (1999) [arXiv:hep-ph/9804281].
13. C. W. Bernard *et al.*, Phys. Rev. D **64**, 054506 (2001) [arXiv:hep-lat/0104002].
14. C. R. Allton *et al.* [UKQCD Collaboration], Phys. Rev. D **65**, 054502 (2002) [arXiv:hep-lat/0107021].

Part III

Computational Methods

A Modular Iterative Solver Package in a Categorical Language

T.J. Ashby[1,2], A.D. Kennedy[2], and M.F.P. O'Boyle[1]

[1] Institute of Computer Systems Architecture, University of Edinburgh
[2] School of Physics, University of Edinburgh
  t.ashby@ed.ac.uk   adk@ph.ed.ac.uk   mob@inf.ed.ac.uk

Summary. Iterative solvers based on Krylov subspace techniques form an important collection of tools for numerical linear algebra. The family of generic solvers is made up of a large number of slightly different algorithms with slightly different properties, which leads to redundancy when implementing them. To overcome this, we build the algorithms out of modular parts, which also makes the connections between the algorithms explicit. In addition to the design of this toolkit, we present here a summary of our initial investigations into automatic compiler optimisations of the code for the algorithms built from these parts. These optimisations are intended to mitigate the inefficiency introduced by modularity.

1 Introduction

Iterative solvers are popular for approximately solving the equation $A\mathbf{x} = \mathbf{b}$ for vector $\mathbf{x}$ given vector $\mathbf{b}$ and square sparse matrix A. They have an advantage over sparse factorisation techniques in that they only rely on being able to perform matrix – vector products and assorted vector operations. Hence there are no problems such as "fill – in" when altering the operator A, which lead to increased storage requirements. For large enough and sparse enough problems, they are the only feasible method.

The iterative solvers based on Krylov subspace techniques form a populous family: FOM, GMRES, IOM(k), CG, SYMMLQ, QMR, etc. These algorithms have much in common, although this is not always obvious from their derivation. Specifically, with the notable exception of the Lanczos-type product methods [4] (LTPMs), they can be formulated using a Lanczos algorithm to generate the basis vectors of the Krylov subspace and the projection of the operator onto that subspace, followed by a decomposition of the projected matrix to find some approximation from the generated subspace. The different combinations of Krylov space generation and decomposition along with the orthogonality condition imposed on the residual at each step give rise to the different algorithms.

On top of the basic flavours of algorithm we can make additions such as multishift solvers, look-ahead to avoid breakdown in the two-sided Lanczos algorithm, etc. The large number of combinations of options means that writing separate codes for the algorithms leads either to repetition in coding or missing out combinations. We solve this problem in our approach by developing a toolkit of pieces that can be plugged together to create the solvers. Building the algorithms from parts can incur some overhead in terms of extra work on an actual machine, however. Explicitly stitching together pieces to form an algorithm presents us with the opportunity to possibly save work that, if we only examined each piece locally, we would not be able to spot. This is where compiler transformations play a part, because the compiler exploits its global knowledge of the program to do the optimisations across modules that may originally have been part of the specific algorithm.

2 Algorithmic Approach

We give here a quick sketch of the algorithms in question to make the basis of our work clear. For a comprehensive exposition of these ideas we refer the reader to [3, 6]. The subject of our attention is the $m \times m$ matrix equation $\mathsf{A}\mathbf{x} = \mathbf{b}$. The basic relationship exploited to construct the solvers is the Arnoldi relation, which describes how to construct a basis of the Krylov space $\mathcal{K}_n(\mathsf{A}, \mathbf{v}_1)$ where $\mathbf{v}_1$ is the first column of the $m \times n$ matrix V_n (where n is the current step):

$$\mathsf{A}\mathsf{V}_n = \mathsf{V}_n \mathsf{H}_{n,n} + \beta_n \mathbf{v}_{n+1} \mathbf{e}_n^H \quad (1)$$
$$= \mathsf{V}_{n+1} \underline{\mathsf{H}}_{n+1,n}$$

This relation is used directly for the long recurrence solvers such as FOM and GM-RES, where V_n has orthonormal columns and $\mathsf{H}_{n,n}$ is upper Hessenberg, both by construction, and A is non-Hermitian. Important simplifications result from either having a Hermitian operator or employing the two-sided Lanczos process to generate biorthogonal matrices V_n and W_n that are non-orthogonal bases of the Krylov spaces $\mathcal{K}_n(\mathsf{A}, \mathbf{v}_1)$ and $\mathcal{K}_n(\mathsf{A}^H, \mathbf{w}_1)$ respectively. The choice of the dual start vector $\mathbf{w}_1$ is arbitrary subject to the constraint that $(\mathbf{v}_1, \mathbf{w}_1) \neq 0$. In both of these cases the Hessenberg matrix in (1) becomes tridiagonal – in the former because the Hermiticity-preserving similarity transformation:

$$\mathsf{V}_n^H \mathsf{A} \mathsf{V}_n = \mathsf{V}_n^H \mathsf{V}_n \mathsf{H}_{n,n} + \beta_n (\mathsf{V}_n^H \mathbf{v}_{n+1}) \mathbf{e}_n^H \quad (2)$$
$$= \mathsf{T}_{n,n}$$

indicates the structure of $\mathsf{H}_{n,n}$, and in the latter case by considering the two two-sided versions of (2) and exploiting the biorthogonality of V_n and W_n. A tridiagonal T implies three-term recurrences to compute the basis V (and W). It is also possible to truncate the long recurrences for a non-Hermitian matrix A (e.g., IOM(k)). This gives a banded upper Hessenberg matrix in (1), but the basis V is no longer orthonormal.

The standard way to proceed is to start from an orthogonality condition on the residual $\mathbf{r}_n = \mathbf{b} - \mathsf{A}\mathbf{x}_n$, and derive a solution in terms of some pivotless decomposition of the projected matrix $\mathsf{H}_{n,n}$ or $\underline{\mathsf{H}}_{n+1,n}$. To specify the orthogonality condition we introduce a matrix S_n and require that $\mathbf{r}_n$ is orthogonal to its columns:

$$S_n^H r_n = S_n^H(b - Ax_n) = 0$$
$$S_n^H AV_n y_n = S_n^H b$$

Here we have explicitly used the fact that the solution x_n is a linear combination of the columns of V_n, with y_n an n vector. The orthogonality condition is usually implicitly tied to the decomposition of the projected Hessenberg matrix that is used in the algorithm. For instance, Galerkin (i.e., orthogonal residual) methods are typically coupled with an LU decomposition, but it is possible to use others[5], such as LQ (which is QR transposed).

The standard groupings of basis generation, orthogonality condition and decomposition with the appropriate name follow. For the Galerkin methods, with an LU decomposition, $S_n = V_n$ gives us FOM or DLanczos (which is closely related to CG) depending on the Hermiticity of A, and $S_n = W_n$ gives us BiDLanczos (related to BiCG). Similarly, for the minimum residual methods with a QR decomposition, $S_n = AV_n$ gives us GMRES or MINRES. We cannot get QMR easily from an orthogonality condition using W because it minimises $\|\beta e_1 - \underline{T}_{n+1,n} y_n\|_2$ directly. By choosing our solution x_n from a different space, $x_n = A^H V_n y_n$ with $V_n \in \mathcal{K}_n(A^H, b)$, we get the minimum error methods:

$$S_n^H AA^H V_n y_n = S_n^H b$$

and choosing $S_n = V_n$ along with an LQ decomposition gives us GMERR or SYMMLQ[3]. These combinations are summarised in Table 1.

Table 1. Summary of popular iterative methods

Name	Basis Generation	Orthogonality Condition	Decomposition
FOM	Arnoldi	Galerkin	LU
GMRES	Arnoldi	Minimum residual	QR
GMERR	Arnoldi	Minimum error	LQ
CG (DLanczos)	Hermitian Lanczos	Galerkin	LU
MINRES	Hermitian Lanczos	Minimum residual	QR
SYMMLQ	Hermitian Lanczos	Minimum error	LQ
BiCG (BiDLanczos)	Two-sided Lanczos	Galerkin	LU
QMR	Two-sided Lanczos	(Quasi) Minimum residual	QR

To get our solution x_n, we substitute the Arnoldi relation into the equation from the orthogonality condition to get an expression in terms of y_n and $H_{n,n}$ or $\underline{H}_{n+1,n}$. We can arrange the RHS to reduce to βe_1 for Galerkin and minimum error, or $\underline{H}_{n+1,n}^H \beta e_1$ for minimum residual. Once we have solved the expression for y_n, we can find x_n as the linear combination of the appropriate basis vectors.

For the long recurrence algorithms we have now finished the derivation. However, for algorithms with short recurrences for generating the Krylov space bases, we also

---

[3] Interestingly there does not seem to be a well known two-sided (quasi) minimum error algorithm.

need a short recurrence for the solution vector to avoid the need to keep or regenerate the basis vectors. We do this by introducing search vectors $\mathbf{p}_n$, also with their own short recurrence. To give the general flavour, we demonstrate for the Galerkin algorithms, using a pivotless LU decomposition in the expression derived from the orthogonality condition:

$$S_n^H A V_n \mathbf{y}_n = S_n^H \mathbf{b}$$
$$T_{n,n} \mathbf{y}_n = \beta \mathbf{e}_1$$
$$L_n U_n \mathbf{y}_n = \beta \mathbf{e}_1$$
$$V_n \mathbf{y}_n = \underbrace{(V_n U_n^{-1})}_{P_n} \underbrace{(L_n^{-1} \beta \mathbf{e}_1)}_{\mathbf{z}_n}$$
$$\mathbf{x}_n =$$

Here $\mathbf{z}_n$ is a vector of length n. Given this definition, we note that as $\mathbf{z}_n$ is defined as the result of a forward substitution, so $\mathbf{z}_n = \mathbf{z}_{n-1} + \zeta_n \mathbf{e}_n$ and hence our short recurrence for $\mathbf{x}$ is:

$$\mathbf{x}_n = P_{n-1} \mathbf{z}_{n-1} + \zeta_n \mathbf{p}_n$$
$$\mathbf{x}_n = \mathbf{x}_{n-1} + \zeta_n \mathbf{p}_n \tag{3}$$

The short recurrence for P_n comes from its definition in terms of V_n and U_n, the latter of which has only a diagonal and superdiagonal term in any column:

$$P_n U_n = V_n$$
$$u_{n,n} \mathbf{p}_n + u_{n-1,n} \mathbf{p}_{n-1} = \mathbf{v}_n$$
$$\mathbf{p}_n = (\mathbf{v}_n - u_{n-1,n} \mathbf{p}_{n-1})/u_{n,n} \tag{4}$$

In a similar fashion, substituting (1) into the orthogonality condition for the minimum residual algorithms yields a least-squares problem which can be solved by a QR decomposition. The R involved now has two terms above the diagonal, so the recurrence for P requires two **p** vectors to be kept. SYMMLQ also uses a QR decomposition, but the grouping leads to an update of the **p** vectors based on the unitary part of the factorisation, Q.

3 Choice of Language

The aim of this work is to factor out as much common structure from the algorithms as possible, and to represent the relationships in sec. 2 actually in the code. As such we needed an environment with excellent support for abstraction, and careful consideration had to be given to the language to use. The following is a summary of some alternatives along with the reasons they were rejected:

- *System of C/C++ macros* (Expression Templates[8], etc.)
  - poor type checking (relies on the underlying language)

- compile time error messages that are difficult to decipher
- very difficult for host language compiler to analyse and optimise
- *3G Languages* (C, FORTRAN, etc.)
  - a fundamental lack of support for abstraction within the language
  - language semantics are frequently an obstacle to optimisation
  - poor type checking
- *Object Oriented Languages* (C++ objects, Java, etc.)
  - inheritance is the *wrong* abstraction for the subset of linear algebra that we are interested in – it is good for subset relations and unary operators but bad for expressing set membership and binary operators
  - a general reliance on dynamic mechanisms, e.g. throwing type-cast exceptions at run-time rather than giving compile-time errors

The final choice was a statically-typed functional language. These languages give type correctness without run-time overheads, good feedback on static errors, and have strong support for abstraction by means of the type system. Specifically, first-class functions and strongly typed functors are a powerful combination. From this family we chose Aldor.

Aldor[2] is a statically-typed mixed functional/imperative language with a powerful type system that includes limited dependent typing. It is designed both for elegance of expression and performance, incorporating a whole program optimisation strategy in its design. The type system has two basic units; *domains* and *categories*. A domain is an abstract data type, i.e., a representation of domain elements and operations on them, and it is typed by belonging to one or more categories. A category is an interface, i.e., a list of function signatures, and categories can inherit from and extend simpler categories. To qualify for membership of a category, a domain has to implement the functions in the interface is some manner. Hence, an algorithm can be specified abstractly in terms of categorical operations. This makes the algorithms both readable and flexible—we can specialise simple domain operations to make them efficient on a given machine without touching the algorithm. For example, parallelism or vectorisation for a high performance computer can be hidden completely within the low level domains; and, less dramatically, generic source routines can be replaced by libraries tuned for performance on a particular workstation (see sec.5).

Aldor is a very small language, so the majority of domains and categories are written by the user. Both domains and categories can be parameterised—a pertinent example of this is a category we have defined called `LinearSpace` which is parameterised by a domain that itself belongs to the category `Field`. In this way we can specify what operations we need on an arbitrary vector space without fixing in advance the domain to represent the ground field.

4 Developing the Framework

The component parts of the solvers are separated and organised using Aldor's advanced type system, and implemented using its functional language features. The collection of pieces comprises a number of different Krylov basis generators based on the different Lanczos processes, different decompositions to apply to the projected matrix, different operators, and more prosaic pieces such as matrices, vectors and scalars. Any given solver that fits into our framework can easily be built from a subset of these pieces.

4.1 Linear Algebra Categories

The linear algebra categories are used to abstract the basic operations that the solvers rely on. An illustrative subset of the types we use are presented in Table 2, where each category is parameterised by a domain from the category listed below it, and a small subset of ancestors that the category inherits from is presented.

Table 2. Examples of Linear Algebra Categories

Category	Sample of ancestors	Example
HermitianLinearOperator	LinearOperator, LinearAlgebra	the matrix A
InnerProductSpace	LinearSpace, Module, Group	vectors **v**, **w**
FieldWithValuation	Ring, EuclideanDomain, ArithmeticSystem	complex scalars α, β
Valuation	Monoid, BasicType	non-negative real scalars γ

The inclusion of the valuation in the hierarchy is interesting as it is possible to use the type system to convey information on the relationships between pieces of the algorithm. For instance, the matrix of normalisation coefficients from the Krylov subspace generated by a Hermitian linear operator on an inner product space consists of scalars from the valuation domain (by Hermiticity and construction) rather than the ground field.

Given that the linear operator is a parameter to the main algorithm, we get versions of the solvers that use pre-conditioning more-or-less for free. For example, to left pre-condition with M, all that is needed is some surrounding code that constructs the appropriately updated right hand side **Mb**, and either explicitly constructs the pre-conditioned operator MA or simply couples the matrices using function composition and wraps them as a single operator. A solver is then called on the new right hand side with the new operator. Right and symmetric pre-conditioning can be treated similarly, but the efficiency gains possible in the special case of symmetric pre-conditioning using some $N = M^2$ would need a special purpose Krylov space recurrence.

4.2 Framework Categories

The framework categories are used to abstract the relationships of parts of the algorithm to one another so that different pieces can be substituted to give different solver algorithms. The high level organisation of the algorithm consists of some initialisation followed by a loop enclosing three nested parts:

- The solution vector update – creating $\mathbf{x}_n$ using $\mathbf{x}_{n-1}$ and $\mathbf{p}_n$

```
V(i) : integer -> vector ==                 step_Forward : () -> () ==
1. compare i with current state j           1.  u := A v1;
2. if i < j then reset state to start       2.  alpha := (u * w1)/delta;
   and recurse with V_column(i)             3.  beta := (gamma * delta)/deltaOld;
3. otherwise do (i - j) calls to            4.  v2 := u - alpha * v1 - beta * v2;
   step_Forward() to cycle the              5.  w2 := AH w1 - conj(alpha) * w1
   recurrence                                         - conj(beta) * w2;
4. update i to equal j                      6.  deltaOld := delta;
5. return the vector v₂                     7.  deltaTemp := (v2 * w2);
                                            8.  gamma := norm(v2);
                                            9.  delta := deltaTemp/gamma^2;
                                            10. (v1, v2) := (v2/gamma, v1);
                                            11. (w1, w2) := (w2/conj(gamma), w1);
```

Fig. 1. Code for the recurrence and pseudo-code for the "lazy matrix" V of a Krylov space basis derived from a two-sided Lanczos process. Initialisation of variables in the recurrence has been omitted.

- The search vector update – creating $\mathbf{p}_n$ with $\mathbf{v}_n$ and some previous number of $\mathbf{p}$ vectors
- The basis vector update – creating $\mathbf{v}_n$ using $A\mathbf{v}_{n-1}$ and some previous $\mathbf{v}$ (and possibly $\mathbf{w}$) vectors

In order to keep as close as possible to the derivation in sec. 2, the search and basis vectors are presented (by means of a category) as a matrix of column vectors (that is P, V and optionally W). The columns of these matrices cannot be enumerated and stored explicitly, so a matrix is implemented lazily with a recurrence to facilitate the illusion. Hence, by way of example, the solution vector update is written in terms of requests for columns of P. These column vectors are generated behind the scenes by the recurrence for the search vectors, which itself makes use of the lazy matrix V.

4.3 Implementing Lazy Matrices

To implement lazy matrices we use Aldor's syntactic sugar for function application to be able to write a column-fetch operation from a matrix as a user function. This function takes some matrix representation and an index as arguments, and gives a vector as a result. The underlying matrix representation is just a (first-class) function, and the column-fetch is a wrapper that applies this function to some integer argument.

The function that represents a matrix in this way generates any given column vector by iterating a recurrence on some hidden state that is set up when the lazy matrix is first constructed. For example, consider a Krylov space object that represents the results of a two-sided Lanczos process. The Krylov space object $\mathcal{K}$ is constructed from an operator A and two start vectors $\mathbf{v}_1$ and $\mathbf{w}_1$. From this somewhat abstract object we derive three lazy matrices – the matrix of basis vectors V, its dual counterpart W, and the tridiagonal matrix of normalisation coefficients T. Each of these is a function from an index value to a column vector, with their respective hidden states initialised with the values A, $\mathbf{v}_1$ and $\mathbf{w}_1$.

A pseudo-code for the lazy matrix V can be found in Fig. 1 along with the actual Aldor code for its underlying recurrence. When this function is called with integer argument i, it cycles the recurrence for the Krylov basis vectors until it has generated the i-th vector, which it then returns. If the sequence of requested vectors is $i, j, k, \ldots$ with $i \leq j \leq k \leq \ldots$, then a procedure that starts from scratch each time will waste effort. A better approach, as demonstrated, is to store the current state after a request i so that we can begin from there next time, provided the next request $j \geq i$. If $j < i$ then we restart. In addition, we can save work and space by sharing the same state with the recurrences for W and T in the hope that the index of any request to one of these matrices is greater than or equal to any previous request to any of them. This of course matches the sequence of indices in the context of iterative solvers, and hence we get correctness and memory economy by laziness and performance by caching.

5 Compiler Optimisations

Separately optimising simple functions where a program spends most of its time can bring large benefits for performance, especially if the pieces are hand coded by experts. However, dividing a program into pieces prevents the sharing of information across module boundaries, and this lack of global information can mean that we miss important opportunities for optimisation. By contrast, a compiler can break the boundaries introduced by modularity and perform cross-component optimisation on a case-by-case basis. The gain from breaking these boundaries may outweigh the benefit of locally superior optimisation in a chain of functions.

By way of simple illustration, consider a pre-compiled library routine for element-wise adding two vectors, and a routine written in source code for the same purpose. If our program needs to add three vectors, then the opaqueness of the library routine forces us to use it twice and make six vector traversals in total (three per call – the two arguments and the result). In contrast, by having the source available a compiler could combine together two copies of the function for adding two vectors to create a specialised function for adding three vectors. This new function would only require four vector traversals in total, by avoiding writing out and then re-reading the intermediate result. This way the code is written using a minimal set of high-level routines but compiled to an efficient specialised program.

We present in this section the experimental results for prototype compiler transformations to exploit cross-component optimisation for our solver codes in this manner, and compare it to a solver augmented with calls to a high performance library. The results show that cross-component optimisation can be highly effective.

5.1 Experiments

Our prototype problem is a nearest neighbour operator on a regular three dimensional grid of complex scalars, such as might arise from approximating a Laplacian using finite differences (i.e. a 3D stencil). We present this as a simple example rather than advocating using this type of solver for these problems. Ultimately we wish to progress to experiments with a pure Wilson-Dirac operator from QCD. The algorithm for the solve is QMR (which uses the code in Fig. 1 to generate its Krylov space basis), and the machine used is a 1GHz Pentium 3 (Coppermine).

Library Augmented Code

This program represents the common library substitutions that a reasonably expert programmer might make. This consists of using the level 1 BLAS functions from a high performance library, the choice here being the ATLAS [9] binary package (version 3.4.1) for the Pentium 3. The sparse matrix-vector product (i.e. the stencil) is not substituted with anything as there is no real standard library that covers all possible sparse matrix schemes. The total number of vector traversals in this version is 24 reads and 12 writes.

Transformed Code

The hand-transformed program is based on aggressive multiple transformations of the pure Aldor version of the code. These transformations include loop fusion, removal of temporaries, constant folding and delaying computations (we refer the reader to [7] for details). The total number of vector traversals in this version is 10 reads and 5 writes, showing a significant increase in locality.

Table 3. Performance of ATLAS augmented against Transformed code for QMR

Problem size	Library augmented (seconds)	Transformed (seconds)	Speed-up of transformed code
10^3	1.15	1.04	1.1
	L1 cache (16kB) $\leq 10^3$		
15^3	4.27	4.27	1.0
20^3	11.87	10.70	1.1
25^3	26.01	20.73	1.25
	$25^3 <$ L2 cache (256kB) $< 26^3$		
30^3	49.86	34.99	1.42
35^3	80.18	60.02	1.33
40^3	136.94	101.00	1.35

Results

Our results are presented in Table 3. They show that, for our example operator, cross-component optimisation can bring large benefits. The library augmented code spends at least 60% percent of its time in the library functions, and at most 40% of its time applying the operator, the (untransformed) code for which is common to both versions.

6 Related Work

There are many iterative solver packages – see [1] for one review. Most packages either prescribe data structures or leave the implementation of low level routines that rely on them to the user. The first of these options necessarily imposes a compromise of efficiency, and the second usually implies a strict separation between the algorithm and the low level routines, limiting optimisation possibilities. Our project differs in that the low level routines are written in Aldor and hence are both customisable and available for ambitious transformations. Some projects based on C++ templates share our unified approach, but they suffer the software engineering drawbacks outlined in sect. 3.

Numerical work in functional languages is rare. The closest piece of work of which we are aware [10] differs by implementing a single iterative method phrased in a non-modular way, and using a language based on normal order evaluation. They do not go into optimisation in any detail.

7 Conclusion and Future Work

In this paper we have presented our approach to engineering a family of numerical algorithms from parts. We chose Aldor as the host language to reduce the software-engineering burden and maximise usability. The choice of language and style of modularisation presents both novel problems and novel opportunities for compiler optimisation which we are investigating.

The current work will be continued to cover more experiments and formalise the transformations. The simplest extensions thereafter would be to add multi-shift solvers and look-ahead. An interesting but more complex extension would be to revise the framework to incorporate LTPMs and eigensolvers.

References

1. Victor Eijkhout. Overview of iterative linear system solver packages. *NHSE review*, 3, 1998.
2. Stephen Watt et al. Aldor. http://www.aldor.org.
3. Anne Greenbaum. *Iterative methods for solving linear systems*. Society for Industrial and Applied Mathematics, 1997.
4. Martin H. Gutknecht and Klaus J. Ressel. Look-ahead procedures for Lanczos-type product methods based on three-term Lanczos recurrences. *SIAM Journal on Matrix Analysis and Applications*, 21(4):1051–1078, 2000.
5. C.C. Paige and M.A. Saunders. Solution of sparse indefinite systems of linear equations. *SIAM Journal Numerical Analysis*, 12:617–629, 1975.
6. Y. Saad. *Iterative Methods for Sparse Linear Systems*. Society for Industrial and Applied Mathematics, 2003.
7. T.J.Ashby, A.D.Kennedy, and M.F.P.O'Boyle. Cross component optimisation in a high level category-based language. Submitted to: EUROPAR 2004.
8. Todd L. Veldhuizen. Expression templates. *C++ Report*, 7(5):26–31, June 1995. Reprinted in C++ Gems, ed. Stanley Lippman.
9. R. Clint Whaley, Antoine Petitet, and Jack J. Dongarra. Automated empirical optimizations of software and the ATLAS project. *Parallel Computing*, 27(1-2):3–35, 2001.
10. Thorsten H.-G. Zörner and Rinus Plasmeijer. Solving linear systems with functional programming languages. In *Implementation of Functional Languages '98*.

Iterative Linear System Solvers with Approximate Matrix-vector Products

Jasper van den Eshof[1], Gerard L.G. Sleijpen[2], and Martin B. van Gijzen[3]

[1] Department of Mathematics, University of Düsseldorf, Universitätsstr. 1, D-40224, Düsseldorf, Germany. `eshof@am.uni-duesseldorf.de`
[2] Department of Mathematics, Utrecht University, P.O. Box 80.010, NL-3508 TA Utrecht, The Netherlands. `sleijpen@math.uu.nl`
[3] CERFACS, 42 Avenue Gaspard Coriolis, 31057 Toulouse Cedex 01, France. `vangijzen@cerfacs.fr`

Summary. There are classes of linear problems for which a matrix-vector product is a time consuming operation because an expensive approximation method is required to compute it to a given accuracy. One important example is simulations in lattice QCD with Neuberger fermions where a matrix multiply requires the product of the matrix sign function of a large sparse matrix times a vector. The recent interest in this and similar type of applications has resulted in research efforts to study the effect of errors in the matrix-vector products on iterative linear system solvers. In this paper we give a very general and abstract discussion on this issue and try to provide insight into why some iterative system solvers are more sensitive than others.

1 Introduction

The central problem in this paper is to find an approximate solution to the equation
$$\mathbf{A}\mathbf{x} = \mathbf{b}.$$

For some linear problems the matrix-vector product can be an expensive operation since a time consuming approximation must be constructed for the product, as for example in simulations in lattice QCD with Neuberger fermions. The recent interest in this, and other applications, has resulted in research efforts to study the impact of errors in the matrix-vector products on iterative linear system solvers, e.g., [3, 4, 8, 10]. The purpose of this paper is to give general and abstract novel insight into why some iterative system solvers are more sensitive than others. This understanding is, for example, important to devise efficient strategies for controlling the errors and, moreover, to choose a suitable iterative solver for a problem. Therefore, we conclude this paper by discussing shortly how this insight can be used to derive strategies for controlling the error. Experiments with these strategies and additional ideas that can be exploited in simulations in lattice QCD with Neuberger fermions are discussed in [1].

2 Krylov Subspace Methods

An important class of iterative solvers for linear systems is the class of *Krylov subspace solvers*. A Krylov subspace method is characterized by the fact that it is an iterative method that constructs its approximate iterate in step j, $\mathbf{x}_j$, from the j dimensional *Krylov subspace*, $\mathcal{K}_j$, defined as the span of $\{\mathbf{b}, \mathbf{A}\mathbf{b}, \ldots, \mathbf{A}^{j-1}\mathbf{b}\}$. There are various ways of constructing these iterates. Of particular importance are Krylov subspace methods that construct their iterates in an optimal way. Two strategies that are considered in this paper are

1. *Galerkin extraction*: where $\mathbf{x}_j = \mathbf{x}_j^{\text{GAL}} \in \mathcal{K}_j$ such that $\mathbf{r}_j^{\text{GAL}} = \mathbf{r}_j := \mathbf{b} - \mathbf{A}\mathbf{x}_j^{\text{GAL}} \perp \mathcal{K}_j$
2. *Minimal residual extraction*: where $\mathbf{x}_j = \mathbf{x}_j^{\text{MR}} \in \mathcal{K}_j$ and $\mathbf{r}_j^{\text{MR}} = \mathbf{r}_j := \mathbf{b} - \mathbf{A}\mathbf{x}_j$ and $\|\mathbf{b} - \mathbf{A}\mathbf{x}_j\|_2$ is minimal.

The observations we make are not difficult to extend to the situation of using other test spaces. Methods like Bi-CGstab [11], however, do not straightforwardly fit into the framework of this paper.

In Krylov subspace methods the Krylov subspace is (implicitly) expanded by applying the matrix to some vector $\mathbf{z}_j$ in step $j + 1$. The vectors $\mathbf{z}_j$ for $j = 0, \ldots, k-1$ necessarily form a basis for the Krylov subspace. (Notice that in this paper we assume that the starting vector of the iterative methods is the zero vector.) We try to provide insight into the influence of (deliberate) errors in the matrix-vector multiplies on the iterative method. It will therefore come as no surprise that the particular choice of the $\mathbf{z}_j$ will play an important role in the remainder of this paper. Other quantities of interest are the *iterates*, $\mathbf{x}_j$, and the *residuals*, $\mathbf{r}_j := \mathbf{b} - \mathbf{A}\mathbf{x}_j$. We will assume that the following relation links together the quantities of interest after k iteration steps:

$$\mathbf{A}\mathbf{Z}_k = \mathbf{R}_{k+1}\underline{S}_k \quad \text{and} \quad \mathbf{x}_k = \mathbf{Z}_k S_k^{-1} e_1, \tag{1}$$

with $\underline{S}_k$ being a $(k+1) \times k$ upper Hessenberg matrix and S_k the $k \times k$ upper block of $\underline{S}_k$. The definition of S_k depends on the method used but we stress that the recursions described by (1) do not have to be *explicitly* used by the particular method. More specifically, the basis used in the extraction of $\mathbf{x}_j$ from the Krylov subspace may differ from the basis of $\mathbf{z}_j$'s used for the expansion. Throughout this paper capital letters are used to group together vectors which are denoted with lower case characters with a subscript that refers to the index of the column (starting with zero for the first column). Hence, $\mathbf{R}_k e_{j+1} = \mathbf{r}_j$.

3 Approximate Matrix-Vector Products

To model the existence of perturbations on the exact matrix-vector products we assume that the matrix-vector products are computed by the function $\mathcal{A}_\eta(\mathbf{v})$ that represents approximations to the matrix-vector product $\mathbf{A}\mathbf{v}$ with a relative precision η as

$$\mathcal{A}_\eta(\mathbf{v}) = \mathbf{A}\mathbf{v} + \mathbf{f} \quad \text{with} \quad \|\mathbf{f}\|_2 \leq \eta \|\mathbf{A}\|_2 \|\mathbf{v}\|_2.$$

The precise source for the existence of these perturbations can be various and are at this point not of interest. We neglect other errors.

In case the matrix-vector product is computed to some relative precision η_j in step $j+1$, we assume that (1) becomes

$$\mathbf{A}\mathbf{Z}_k + \mathbf{F}_k = \mathbf{R}_{k+1}\underline{S}_k \quad \text{and} \quad \mathbf{x}_k = \mathbf{Z}_k S_k^{-1} e_1. \tag{2}$$

The vector $\mathbf{f}_j$ is the $j+1$-th column of $\mathbf{F}_k$ and it contains the error in the matrix-vector product in step $j+1$ and we, therefore, have that $\|\mathbf{f}_j\|_2 \leq \eta_j \|\mathbf{A}\|_2 \|\mathbf{z}_j\|_2$. It can be easily checked that this assumption is appropriate for all inexact Krylov methods that we consider in this paper. (Notice that we assume that there are no roundoff errors.) The perturbation $\mathbf{F}_k$ in (2) causes that $\mathbf{r}_k$ is not a residual for the vector $\mathbf{x}_k$ defined by the second relation. As a consequence one should be careful when assessing the accuracy of the iterate $\mathbf{x}_k$. Instead we have the following inequality involving the norm of the *residual gap*, that is the distance between the *true residual*, $\mathbf{b} - \mathbf{A}\mathbf{x}_k$, and the *computed residual*, $\mathbf{r}_k$:

$$\|\underbrace{\mathbf{b} - \mathbf{A}\mathbf{x}_k}_{\text{true residual}}\|_2 \leq \|\underbrace{\mathbf{r}_k - (\mathbf{b} - \mathbf{A}\mathbf{x}_k)}_{\text{residual gap}}\|_2 + \|\underbrace{\mathbf{r}_k}_{\text{computed residual}}\|_2.$$

We notice that the size of the true residual is unknown in contrast to the size of the computed residual and it follows from (2) that the gap is bounded by

$$\|\mathbf{r}_k - (\mathbf{b} - \mathbf{A}\mathbf{x}_k)\|_2 = \|\mathbf{F}_k S_k^{-1} e_1\|_2 \leq \sum_{j=0}^{k-1} \eta_j \|\mathbf{A}\|_2 \|\mathbf{z}_j\|_2 |e_{j+1}^* S_k^{-1} e_1|. \tag{3}$$

Focusing on the residual gap is not uncommon in theoretical analyses of the attainable accuracy of iterative methods in the finite precision context, see e.g., [9]. It is based on the frequent observation that the computed residuals eventually become many orders of magnitude smaller than machine precision and, therefore, the attainable precision is determined by the size of the residual gap. A similar technique can be used for getting insight into the effect of approximate matrix-vector products on Krylov methods: if we terminate as soon as $\|\mathbf{r}_k\|_2$ is of order ϵ, then the size of the gap determines the precision of the inexact process. When we plot the convergence curve of the true residuals then from some point on the decrease of the residual norms stagnates and the level at which this occurs is determined by the size of the residual gap. Extensive numerical experiments confirm this observation for other problems. With this assumption we see that the sensitivity of the stagnation level of a particular Krylov subspace method is determined by the quantities $\|\mathbf{z}_j\|_2 |e_{j+1}^* S_k^{-1} e_1|$. The purpose is to investigate the size of these quantities to give some understanding of the sources that influence the sensitivity of Krylov subspace methods.

Some preliminary insight can be given. We have that

$$\underbrace{\mathbf{x}_k}_{\text{extraction}} = \underbrace{\mathbf{Z}_k}_{\text{choice basis}} S_k^{-1} e_1.$$

This shows that the size of the elements of this vector do not only depend on the optimality properties of the iterates (i.e., how $\mathbf{x}_k$ is chosen from $\mathcal{K}_k$) but also on the choice of the basis given by the $\mathbf{z}_j$. On termination, when $\mathbf{x}_k \approx \mathbf{x}$, we expect no essential difference between a Galerkin extraction and minimal residual extraction. On the other hand, linear dependence in the matrix $\mathbf{Z}_k$ implies that elements of the vectors $|S_k^{-1} e_1|$ (where the absolute values are taken elementwise) can be relatively large and results in a large sensitivity of the particular method to inexact matrix-vector products. The choice of the expansion basis, rather than the extraction method, determines the sensitivity to errors in the matrix-vector products. In the following we will make this statement more precise.

3.1 The General Case

We study the size of the elements of $\|\mathbf{z}_j\|_2 |e_{j+1}^* S_k^{-1} e_1|$ by assuming *exact* matrix-vector products for the moment, i.e., (1) holds. This problem was studied in related formulation in [8, 10]. We first have to introduce some notation. Let $\mathbf{M}$ and $\mathbf{N}$ be Hermitian, positive definite, n dimensional matrices. We define

$$\delta_{\mathbf{M} \to \mathbf{N}} \equiv \max_{\mathbf{y} \neq \mathbf{0}} \frac{\|\mathbf{y}\|_{\mathbf{M}}}{\|\mathbf{y}\|_{\mathbf{N}}}$$

which gives the following norm equivalence

$$(\delta_{\mathbf{N} \to \mathbf{M}})^{-1} \|\mathbf{y}\|_{\mathbf{N}} \leq \|\mathbf{y}\|_{\mathbf{M}} \leq \delta_{\mathbf{M} \to \mathbf{N}} \|\mathbf{y}\|_{\mathbf{N}}. \tag{4}$$

We furthermore define the inner product $<\mathbf{z}, \mathbf{y}>_{\mathbf{M}} \equiv \mathbf{z}^* \mathbf{M} \mathbf{y}$ and assume that $\mathbf{Z}_k$ is an $\mathbf{M}$-orthogonal basis (that is, the columns of the matrix $\mathbf{Z}_k$ are orthogonal in the $<\cdot,\cdot>_{\mathbf{M}}$ inner product). Now we have for all $\tilde{\mathbf{x}}_j \in \mathcal{K}_j$

$$\begin{aligned} |e_{j+1}^* S_k^{-1} e_1| \|\mathbf{z}_j\|_{\mathbf{M}}^2 &= |<\mathbf{z}_j, \mathbf{x}_k>_{\mathbf{M}}| \\ &= |<\mathbf{z}_j, \mathbf{x}_k - \tilde{\mathbf{x}}_j>_{\mathbf{M}}| \leq \|\mathbf{x}_k - \tilde{\mathbf{x}}_j\|_{\mathbf{M}} \|\mathbf{z}_j\|_{\mathbf{M}}. \end{aligned} \tag{5}$$

Here we have made use of the fact that $<\mathbf{z}_j, \tilde{\mathbf{x}}_j>_{\mathbf{M}} = 0$. Recall that $\mathbf{x}_j^{\text{MR}}$ is defined as the approximation from the space $\mathcal{K}_j$ that minimizes the error

Table 1. Values for various Krylov subspace methods assuming that $\mathbf{M} = \mathbf{M}^*$ and $\mathbf{N} = \mathbf{N}^*$ are strictly positive definite.

Example method	$\mathbf{M}$	$\mathbf{N}$	$\delta_{\mathbf{I} \to \mathbf{M}}$	$\delta_{\mathbf{M} \to \mathbf{N}}$	$\delta_{\mathbf{N} \to \mathbf{A}^*\mathbf{A}}$
ORTHORES	$\mathbf{I}$	$\mathbf{A}$	1	$\sqrt{\|\mathbf{A}^{-1}\|_2}$	$\sqrt{\|\mathbf{A}^{-1}\|_2}$
GMRES	$\mathbf{I}$	$\mathbf{A}^*\mathbf{A}$	1	$\|\mathbf{A}^{-1}\|_2$	1
CG	$\mathbf{A}$	$\mathbf{A}$	$\sqrt{\|\mathbf{A}^{-1}\|_2}$	1	$\sqrt{\|\mathbf{A}^{-1}\|_2}$
GCR	$\mathbf{A}$	$\mathbf{A}^*\mathbf{A}$	$\sqrt{\|\mathbf{A}^{-1}\|_2}$	$\sqrt{\|\mathbf{A}^{-1}\|_2}$	1

in $\mathbf{A}^*\mathbf{A}$-norm, or, equivalently, minimizes the 2-norm of the residual $\mathbf{r}_j^{\mathrm{MR}} = \mathbf{b} - \mathbf{A}\mathbf{x}_j^{\mathrm{MR}}$. With this definition and (5), we get the bound

$$\|\mathbf{z}_j\|_2 |e_{j+1}^* S_k^{-1} e_1| \leq \frac{\|\mathbf{z}_j\|_2}{\|\mathbf{z}_j\|_{\mathbf{M}}} \left(\|\mathbf{x} - \mathbf{x}_j^{\mathrm{MR}}\|_{\mathbf{M}} + \|\mathbf{x} - \mathbf{x}_k\|_{\mathbf{M}} \right)$$
$$\leq \delta_{\mathbf{I} \to \mathbf{M}} \, \delta_{\mathbf{M} \to \mathbf{A}^*\mathbf{A}} \left(\|\mathbf{r}_j^{\mathrm{MR}}\|_2 + \|\mathbf{r}_k\|_2 \right). \quad (6)$$

This simple argument in combination with the bound on the residual gap (3) suggests that, if the inexact Krylov subspace method is terminated as soon as $\|\mathbf{r}_k\|_2 \leq \epsilon$, then the size of the residual gap is essentially bounded by a constant times the norm of the residuals corresponding to a minimal residual extraction. Notice that these residuals form a monotonically decreasing sequence and are therefore bounded.

If, for some $\mathbf{N}$, the particular Krylov subspace method produces an iterate, $\tilde{\mathbf{x}}_j$, that minimizes the error in the $\mathbf{N}$-inner product, then we can even remove the $\|\mathbf{r}_k\|_2$ term in (6): we have that $\|\mathbf{x} - \tilde{\mathbf{x}}_j\|_{\mathbf{N}}^2 = \|\mathbf{x}_k - \tilde{\mathbf{x}}_j\|_{\mathbf{N}}^2 + \|\mathbf{x}_k - \mathbf{x}\|_{\mathbf{N}}^2$. Using this we get

$$\|\mathbf{x}_k - \tilde{\mathbf{x}}_j\|_{\mathbf{M}} \leq \delta_{\mathbf{M} \to \mathbf{N}} \|\mathbf{x}_k - \tilde{\mathbf{x}}_j\|_{\mathbf{N}} \leq \delta_{\mathbf{M} \to \mathbf{N}} \|\mathbf{x} - \tilde{\mathbf{x}}_j\|_{\mathbf{N}} \leq \delta_{\mathbf{M} \to \mathbf{N}} \, \delta_{\mathbf{N} \to \mathbf{A}^*\mathbf{A}} \|\mathbf{r}_j^{\mathrm{MR}}\|_2,$$

which leads to the bound

$$\|\mathbf{z}_j\|_2 |e_{j+1}^* S_k^{-1} e_1| \leq \delta_{\mathbf{I} \to \mathbf{M}} \, \delta_{\mathbf{M} \to \mathbf{N}} \, \delta_{\mathbf{N} \to \mathbf{A}^*\mathbf{A}} \|\mathbf{r}_j^{\mathrm{MR}}\|_2. \quad (7)$$

For several well-known Krylov subspace methods we have summarized the relevant quantities in Table 1. Substituting these values into (7) finally shows, for all methods mentioned in the table, that

$$\|\mathbf{z}_j\|_2 |e_{j+1}^* S_k^{-1} e_1| \leq \|\mathbf{A}^{-1}\|_2 \|\mathbf{r}_j^{\mathrm{MR}}\|_2. \quad (8)$$

From our discussion it is clear that the optimality properties of the iterates can simplify the bound (6) somewhat. Since we terminate as soon as $\|\mathbf{r}_k\|_2 \leq \epsilon$, it follows that the impact of the choice of the optimality properties (i.e., the $\mathbf{N}$-inner product) for the iterates is small. (However, it can be large during some iteration steps of the iterative process.) A more important factor in the sensitivity for approximate matrix-vector products is the conditioning of the basis $\mathbf{z}_0, \ldots, \mathbf{z}_{k-1}$ which is determined by the choice of the matrix $\mathbf{M}$. For example, if $\mathbf{M} = \mathbf{A}$ and $\mathbf{A}$ is indefinite then the basis can be almost linear dependent. See [10] for a different point of view, analysis and examples. We will focus on this in the next section

3.2 The Matrix $\mathbf{Z}_k$ is A-orthogonal

In this section we consider the situation that $\mathbf{z}_i^* \mathbf{A} \mathbf{z}_j = 0$ for $i < j$. Or in other words, $\mathbf{Z}_k^* \mathbf{A} \mathbf{Z}_k$ is upper triangular which reduces to a diagonal matrix in case $\mathbf{A}$ is Hermitian. If the matrix $\mathbf{A}$ is Hermitian positive definite then the vectors $\mathbf{z}_j$ form an orthogonal basis with respect to the $\mathbf{A}$-inner product and, therefore, $\mathbf{Z}_k$ is orthogonal with respect to a well-defined inner product. Consequently, we do not expect that the elements of the vector $\|\mathbf{z}_j\|_2 |S_k^{-1} e_1|$ can be arbitrary large as is shown by equation (8) in the previous section.

For general matrices $\mathbf{A}$, the situation is more problematic. Assuming that $\mathbf{z}_j = \mathbf{r}_j^{\text{MR}}$, we prove in the appendix for minimal residual extraction the following, reasonably sharp, estimate:

$$|e_{j+1}^* S_k^{-1} e_1| \leq \|\mathbf{A}^{-1}\|_2 \left(\frac{\|\mathbf{r}_j^{\text{GAL}}\|_2}{\|\mathbf{r}_j^{\text{MR}}\|_2} + \frac{\|\mathbf{r}_{j+1}^{\text{GAL}}\|_2}{\|\mathbf{r}_j^{\text{MR}}\|_2} \frac{\|\mathbf{r}_{j+1}^{\text{MR}}\|_2}{\|\mathbf{r}_j^{\text{MR}}\|_2} \right). \qquad (9)$$

This shows that the j-th element of the vector $|S_k^{-1} e_1|$ might be large if the Galerkin process has a very large residual in the $j-1$-th or j-th step. This reflects near linear dependence in the columns of the matrix $\mathbf{Z}_k$ and results in relatively large upper bound on the residual gap.

To illustrate the previous observations, we have included the results of a simple numerical experiment in Figure 1 where the matrix is diagonal with elements $\{1, 2, \ldots, 100\} - 5.2025$ and the right-hand side has all components equal. The matrix is constructed such that the Galerkin residual is very large in the fifth step. As a result the $\mathbf{A}$-orthogonal basis is ill conditioned. Approximate matrix-vector products are simulated by adding random vectors of relative size 10^{-10} to the exact product. For the methods mentioned in Table 1 we have included the results in this figure.

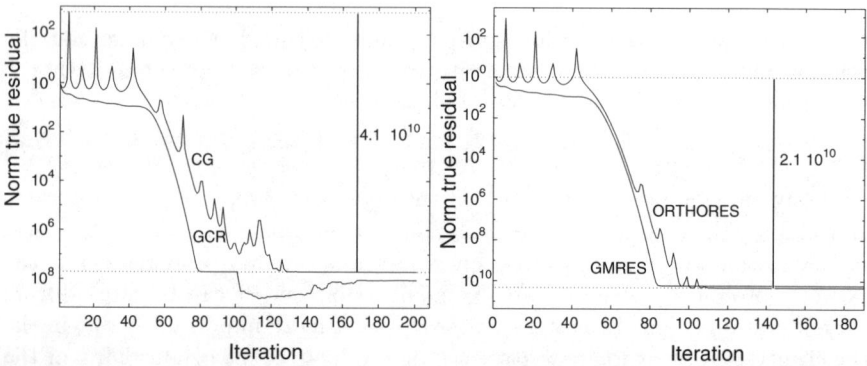

Fig. 1. Illustration that the use of "$\mathbf{A}$-orthogonal" vectors $\mathbf{z}_j$ can have a negative impact on the stagnation level of a method in case of very large intermediate Galerkin residuals.

The left picture shows the convergence curves of the true residuals of the CG and GCR methods. Both methods apply their matrix-vector products to, in the exact case, an **A**-orthogonal basis. As predicted by (9) and the expression for the gap (3), we see that the height of the largest peak of the Galerkin residuals (which coincide with the CG residuals) determines the precision that can be achieved with the GCR method. The stagnation level for the CG method is not very different, as expected. In the right picture we have two methods, ORTHORES and GMRES, that apply their matrix-vector products to an, in the exact case, orthogonal basis. The height of the largest peak is here not of importance and, in fact, the attainable precision is close to 10^{-10}. Notice that the choice of the extraction technique is not of influence on the stagnation level. Both GCR and GMRES employ minimal residual extraction, while CG and ORTHORES use the Galerkin approach. The use of short recurrences does not play a role either. In contrast to GMRES, ORTHORES relies on short recurrences.

4 Discussion

In the previous section we considered the size of the quantities $\|\mathbf{z}_j\|_2 |e_{j+1}^* S_k^{-1} e_1|$ in exact iterative methods. This allowed us to give an abstract and general discussion and identify the main sources of sensitivity towards errors in the matrix-vector products. This allows us to take a fresh point of view on stability issues in iterative linear system solvers. For example, in rounding error analyses of the *conjugate gradient* method, traditionally the sensitivity of this method in case of a large intermediate residual is attributed to instabilities in the Cholesky decomposition implicitly made in the CG method, e.g., [2]. Here we argue that it can be explained by the fact that we work with an ill-conditioned basis. This also explains why we see precisely the same stagnation level in the GCR method, which does not involve a Cholesky decomposition and, moreover, uses full orthogonalization and extracts its iterates such that they are minimal residual approximations, in contrast to CG. Moreover, our heuristic framework also gives an alternative explanation for the observations in [6] that the impact of rounding errors on the attainable accuracy of Krylov methods is not essentially influenced by the smoothness of the residual convergence curve. Notice, that possible instabilities are caused by the choice of an inappropriate solution method and are not part of the problem to be solved itself and can be easily circumvented by switching to a different method. For example, instead of CG for indefinite problems one can use ORTHORES.

We considered the size of the quantities in the exact case. Nevertheless, this is in some sense a best case scenario: if these elements are large then, they are not expected to be small in the practical case. Moreover, they give a good understanding of what we see happen in practice and certainly provides a good guideline for the selection of a suitable Krylov subspace as a solver. For most methods in case of perturbed matrix-vector products, i.e., we are in the situation of (2), analogous results can be derived by interpreting the vector $S_k^{-1} e_1$ as constructed by an exact process applied to a Hessenberg matrix with starting approximate solution vector e_1 which then proves the bound that is given in [10, 8] for the inexact GMRES method.

5 Practical Consequences: Relax to the Max

The previous sections we gave insight into the effect of approximate matrix-vector products on Krylov subspace solvers for linear systems. An important practical consequence of this work should be the construction of efficient strategies for controlling the error in the products such that the overall cost of approximating the products is as small as possible. From a practical point of view this means that we should allow the errors in the matrix-vector products to be as large as possible. In [10] strategies for choosing the η_j are derived by bounding each summand of the sum in (3) on a small, appropriate multiple of ϵ. Combining this remark with Equation (8) suggests, in the relevant circumstances, to use a relative precision for the matrix-vector product in step $j+1$ that is bounded by

$$\eta_j = \frac{\epsilon}{\|\mathbf{r}_j^{\text{MR}}\|_2}. \tag{10}$$

In some iterative methods we have only available the length of $\mathbf{r}_j^{\text{GAL}}$ instead of $\mathbf{r}_j^{\text{MR}}$, as for example the CG method. However, we notice that (see, e.g., [5])

$$\|\mathbf{r}_j^{\text{MR}}\|_2 = \left(\sum_{i=0}^{j} \|\mathbf{r}_i^{\text{GAL}}\|_2^{-2} \right)^{-1/2}.$$

An interesting property of (10) is that it requires very accurate matrix-vector products in the beginning of the process, and the precision is relaxed as soon as the method starts converging, that is, the residuals become increasingly smaller. This justifies the term *relaxation strategy* for this choice of η_j which was introduced by Bouras and Frayssé who reported various numerical results for the GMRES method in [3]. For an impressive list of numerical experiments they observe that the GMRES method with tolerance (10) converges roughly as fast as the unperturbed version, despite the, sometimes large, perturbations. Furthermore, the norm of the true residual ($\|\mathbf{b} - \mathbf{A}\mathbf{x}_j\|_2$) seemed to stagnate around a value of $\mathcal{O}(\epsilon)$.

Despite the recent efforts, the theoretical understanding of the effect of perturbations of the matrix-vector products is still not complete, see for more discussion and references e.g., [10, 7]. In particular the effect of perturbations on the convergence speed is not yet fully understood. (Practical experience with these strategies is however very promising: the speed of convergence does not seem to be much affected.) We note, however, that the convergence speed can be cheaply monitored during the iteration process, whereas the residual gap can only be computed using an expensive matrix-vector product.

Acknowledgments

We would like to thank the referees for their perceptive comments that helped us to improve the presentation of this paper. The first author wishes to thank the organizers of 'The third international workshop on QCD and Numerical Analysis' for their invitation and hospitality in Edinburgh. Part of the work of J. van den Eshof was financially supported by the Dutch scientific organization (NWO) through project 613.002.035.

A A Technical Result

For convenience we assume in this appendix that a vector is appended with additional zeros if this is necessary to make dimensions match. In this appendix we prove (9). In its proof we need the *Arnoldi relation*

$$\mathbf{A}\mathbf{V}_k = \mathbf{V}_{k+1}\underline{T}_k, \text{ with } \mathbf{V}_k e_1 = \mathbf{b},$$

where $\mathbf{V}_k$ spans the k dimensional Krylov subspace $\mathcal{K}_k$ and $\underline{T}_k$ is $k+1 \times k$ upper Hessenberg. We recall that the minimal residuals are equal to the GMRES residuals given our assumption of exact arithmetic and matrix-vector products. Assume that $\underline{T}_k$ has full rank and define the vector $\overline{\gamma}_k = (\gamma_0, \ldots, \gamma_k)^* \in \mathbb{R}^{k+1}$ such that $\overline{\gamma}_k^* \underline{T}_k = \overline{0}^*$ and $e_1^* \overline{\gamma}_k = 1$. It was shown in [10] that

$$\mathbf{r}_k^{\mathrm{MR}} = \|\overline{\gamma}_k\|_2^{-2}\mathbf{V}_{k+1}\overline{\gamma}_k \quad \text{and} \quad \mathbf{r}_k^{\mathrm{GAL}} = \gamma_k^{-1}\mathbf{V}_{k+1}e_{k+1}. \tag{11}$$

The relation between the vector $\overline{\gamma}_k$ and the residuals can be expressed for the residuals $\mathbf{r}_j^{\mathrm{MR}}$ with $j = 0, \ldots, k-1$ by the relation

$$\mathbf{R}_k^{\mathrm{MR}} = \mathbf{V}_k \Upsilon_k \Theta_k^{-2} \quad \text{with} \quad \Upsilon_k e_{j+1} = \overline{\gamma}_j \text{ and } \Theta_k = \mathrm{diag}(\|\overline{\gamma}_0\|_2, \ldots, \|\overline{\gamma}_{k-1}\|_2).$$

This gives us a QR-decomposition for the matrix $\mathbf{R}_k^{\mathrm{MR}}$ which we need in the following lemma.

Lemma 1. *For exact minimal residual approximations with* $\mathbf{Z}_k = \mathbf{R}_k^{MR}$ *and without preconditioning, we have that*

$$S_k^{-1}e_1 = (\mathbf{A}\mathbf{R}_k^{MR})^\dagger \mathbf{r}_0^{MR} = \Theta_k^2 \Upsilon_k^{-1} \underline{T}_k^\dagger e_1 \tag{12}$$

$$e_{j+1}^* \Theta_k^2 \Upsilon_k^{-1} = \frac{\|\overline{\gamma}_j\|_2^2}{\gamma_j} e_{j+1}^* - \frac{\|\overline{\gamma}_j\|_2^2}{\gamma_{j+1}} e_{j+2}^*. \tag{13}$$

Here, $\mathbf{M}^\dagger$ denotes the Moore-Penrose generalized inverse of a matrix $\mathbf{M}$.

Proof. As observed we have that $\mathbf{A}\mathbf{R}_k^{\mathrm{MR}} = \mathbf{A}\mathbf{V}_k\Upsilon_k\Theta_k^{-2}$. This gives

$$(\mathbf{A}\mathbf{R}_k^{\mathrm{MR}})^\dagger \mathbf{r}_0^{\mathrm{MR}} = (\mathbf{A}\mathbf{V}_k\Upsilon_k\Theta_k^{-2})^\dagger \mathbf{r}_0^{\mathrm{MR}} = \Theta_k^2\Upsilon_k^{-1}(\mathbf{A}\mathbf{V}_k)^\dagger \mathbf{r}_0^{\mathrm{MR}} = \Theta_k^2\Upsilon_k^{-1}\underline{T}_k^\dagger e_1.$$

Equality (13) follows from the observation that $\Upsilon_k = \mathrm{diag}(\overline{\gamma}_{k-1})J_k^{-1}$ where J_k is lower bidiagonal with -1 on its subdiagonal and 1 on its diagonal.

To bound the elements of the vector $\underline{T}_k^\dagger e_1$ we can use the observation that the Hessenberg matrix $\underline{T}_k$ is equal to the generated Hessenberg matrix for an exact GMRES process applied to the matrix $\underline{T}_k$ with starting vector e_1. Now we can use, with some additional work, the presented bounds in Section 3.1 (or the equivalent ones from [10, 8] for the GMRES method). In combination with (12) and (13) this gives, for general matrices $\mathbf{A}$,

$$|e_{j+1}^* S_k^{-1} e_1| \leq \|\underline{T}_k^\dagger\|_2 \left(\frac{\|\overline{\gamma}_j\|_2}{|\gamma_j|} + \frac{\|\overline{\gamma}_j\|_2}{|\gamma_{j+1}|} \frac{\|\overline{\gamma}_j\|_2}{\|\overline{\gamma}_{j+1}\|_2} \right)$$

$$\leq \|\mathbf{A}^{-1}\|_2 \left(\frac{\|\mathbf{r}_j^{\text{GAL}}\|_2}{\|\mathbf{r}_j^{\text{MR}}\|_2} + \frac{\|\mathbf{r}_{j+1}^{\text{GAL}}\|_2}{\|\mathbf{r}_j^{\text{MR}}\|_2} \frac{\|\mathbf{r}_{j+1}^{\text{MR}}\|_2}{\|\mathbf{r}_j^{\text{MR}}\|_2} \right).$$

In the last inequality we have used (11).

References

1. G. Arnold, N. Cundy, J. van den Eshof, A. Frommer, S. Krieg, Th. Lippert, and K. Schäfer. Numerical methods for the QCD overlap operator: III. Nested iterations. *Computer Physics Communications*, 2004. In Press.
2. R. E. Bank and T. F. Chan. A composite step bi-conjugate gradient algorithm for nonsymmetric linear systems. *Numer. Algorithms*, 7(1):1–16, 1994.
3. A. Bouras and V. Frayssé. A relaxation strategy for inexact matrix-vector products for Krylov methods. Technical Report TR/PA/00/15, CERFACS, France, 2000.
4. A. Bouras, V. Frayssé, and L. Giraud. A relaxation strategy for inner-outer linear solvers in domain decomposition methods. Technical Report TR/PA/00/17, CERFACS, France, 2000.
5. P. N. Brown. A theoretical comparison of the Arnoldi and GMRES algorithms. *SIAM J. Sci. Stat. Comput.*, 12(1):58–78, 1991.
6. M. H. Gutknecht and M. Rozložnik. Residual smoothing techniques: do they improve the limiting accuracy of iterative solvers? *BIT*, 41(1):86–114, 2001.
7. V. Simoncini and D. B. Szyld. On the superlinear convergence of exact and inexact Krylov subspace methods. Technical report, 2003. To appear in SIAM Review.
8. V. Simoncini and D. B. Szyld. Theory of inexact Krylov subspace methods and applications to scientific computing. *SIAM J. Sci. Comput.*, 25(2):454–477, 2003.
9. G. L. G. Sleijpen, H. A. van der Vorst, and D. R. Fokkema. BiCGstab(ℓ) and other hybrid Bi-CG methods. *Numer. Algorithms*, 7(1):75–109, 1994.
10. J. Van den Eshof and G. L. G. Sleijpen. Inexact Krylov subspace methods for linear systems. *SIAM J. Matrix Anal. Appl.*, 26(1):125–153, 2004.
11. H. A. van der Vorst. Bi-CGSTAB: A fast and smoothly converging variant of Bi-CG for the solution of nonsymmetric linear systems. *SIAM J. Sci. Stat. Comput.*, 13(2):631–644, 1992.

What Can Lattice QCD Theorists Learn from NMR Spectroscopists?

George T. Fleming[1]

Jefferson Lab, 12000 Jefferson Ave, Newport News VA 23606, USA
flemingg@jlab.org

Summary. The Lattice QCD (LQCD) community has occasionally gone through periods of self-examination of its data analysis methods and compared them with methods used in other disciplines [22, 16, 18]. This process has shown that the techniques widely used elsewhere may also be useful in analyzing LQCD data. It seems that we are in such a period now with many groups trying what are generally called Bayesian methods such as Maximal Entropy (MEM) or constrained fitting [19, 15, 1, 7, 5, and many others]. In these proceedings we will attempt to apply this process to a comparison of data modeling techniques used in LQCD and NMR Spectroscopy to see if there are methods which may also be useful when applied to LQCD data.

1 Lattice QCD and NMR Spectroscopy

A common problem in Lattice QCD is the estimation of hadronic energies $E_k(\mathbf{p})$ of $k = 1 \cdots K$ states from samples of the hadronic correlation function of a specified set of quantum numbers computed in a Monte Carlo simulation. A typical model function is

$$C(\mathbf{p}, t_n) = \sum_{k=1}^{K} A_k(\mathbf{p}) \exp\left[-(t_0 + na)E_k(\mathbf{p})\right] \quad (1)$$

$$A_k, E_k \in \mathbb{R}, \quad 0 \leq E_1 \leq \cdots \leq E_k \leq E_{k+1} \leq \cdots \leq E_K$$

where one of the quantum numbers, the spatial momentum $\mathbf{p}$, is shown explicitly for illustration. The correlation function is estimated at each time t_n, $n = 0 \cdots N - 1$ with the N chosen such that $(E_2 - E_1)t_{N-1} \gg 1$. This enables the ground state energy E_1 to be easily determined from the large time behavior. To accurately estimate the k-th energy level requires choosing a sampling interval $a^{-1} \gg E_k - E_1$. Unfortunately, computational constraints typically force us to choose time intervals larger ($a^{-1} \sim 2$ GeV) and number of time samples smaller ($N \sim 32$) than is ideally preferred.

In an idealized nuclear magnetic resonance (NMR) spectroscopy[1] experiment, a sample is placed in an external magnetic field and a transient field from an RF coil is used to temporarily drive the various nuclei into a non-equilibrium distribution of magnetic spin states. Then, as the sample relaxes back to its equilibrium distribution, each type of excited nuclei radiates at a characteristic frequency f_k. The sum of all these microscopic signals are picked up by another RF coil, giving rise to the free induction decay (FID) signal

$$y_n = \sum_{k=1}^{K} a_k e^{i\phi_k} e^{(-d_k + i2\pi f_k)t_n} + e_n, \quad n \in [0, N-1] \qquad (2)$$

$$a_k, \ \phi_k, \ d_k, \ f_k \in \mathbb{R}; \quad d_k \geq 0; \quad \text{noise}: e_n \in \mathbb{C}.$$

As the frequencies are known *a priori*, an experienced operator can incorporate this prior knowledge by Fourier transforming the data and matching Lorentzian peaks against existing databases. Bayesian methods are then used to constrain the frequencies, enabling the estimation of the other parameters. Of particular interest are the amplitudes a_k, which are related to the number of various nuclei in the sample, and the damping rates d_k, which are related to the mobility and molecular environment of the nuclei.

Both Eqs. (1) and (2) can be written in the form

$$y_n = \sum_{k=1}^{K} a_k \alpha_k^n \qquad (3)$$

or in matrix notation $\mathbf{y} = \mathbf{\Phi}(\boldsymbol{\alpha}) \mathbf{a}$. In numerical analysis, this is known as a Vandermonde system and $\mathbf{\Phi}$ is a Vandermonde matrix. Note also that all the parameters $\boldsymbol{\alpha}$ which enter non-linearly in the model only appear in the Vandermonde matrix and the remaining linear parameters in $\mathbf{a}$. This suggests that if the best fit values of only the non-linear parameters, $\widehat{\boldsymbol{\alpha}}$, were known *a priori* then the remaining best fit values of the linear parameters, $\widehat{\mathbf{a}}$ could be determined using a linear least squares algorithm. Hence, linear and non-linear parameters need not be determined simultaneously and in Sec. 2 we will discuss the best known algorithm that exploits this feature.

We have found that all of the model functions we use to fit hadronic correlations in LQCD can be written in the Vandermonde form. For a less trivial example, here is the model function for mesonic correlations with periodic (or anti-periodic) temporal boundary conditions and either Wilson (σ=1) or staggered (σ=-1) fermions

$$C(\tau_n) = \sum_{k=1}^{K} \sigma^{kn} A_k e^{-aNE_k/2} \cosh(anE_k), \quad 0 \leq E_k \leq E_{k+2}.$$

In this case, if we choose $\alpha_k = \sigma^k \cosh(aE_k)$ to be the parameters of the Vandermonde matrix $\mathbf{\Phi}$ then we can construct the data vector $\mathbf{y}$ from the correlation data

$$y_n = \frac{1}{2^{n-1}} \sum_{j=0}^{n-1} \binom{n-1}{j} C(\tau_{n-2j-1}).$$

---

[1] In medical applications, MRS is a preferred abbreviation, probably to avoid the perceived public aversion to anything *nuclear*.

where $\binom{n}{j}$ are binomial coefficients.

In NMR spectroscopy and in LQCD, fitting data often requires an experienced user to interact with the fitting program, *i.e.* to provide initial guesses to the minimizer or to choose what prior knowledge may be used to constrain the minimization, and this can often be a time-consuming process if the data are of marginal quality. In LQCD fitting programs, the *effective mass* technique is often used to provide non-interactive initial guesses to the minimizer. In NMR spectroscopy, more general analogues, called *black box* methods, have been developed for situations where an expert user is unavailable or the rate of data acquisition precludes interaction. In Sec. 3, we will look at the generalization of the effective mass technique, which will lead to a Hankel system that must be solved.

2 VARPRO: Variable Projection Algorithm

In Sec. 1, we considered data whose model may be written as $\mathbf{y} = \mathbf{\Phi}\mathbf{a}$, as in Eq. (3), with the data vector $\mathbf{y} \in \mathbb{R}^N$ and the linear parameter vector $\mathbf{a} \in \mathbb{R}^K$ and $N > 2K$ is necessary for the problem to be over-determined. The non-linear parameter vector $\boldsymbol{\alpha}$ is used to determine the components of the non-linear parameter matrix $\mathbf{\Phi} \in \mathbb{R}^{N \times K}$ of the general form

$$\mathbf{\Phi} = \begin{pmatrix} \phi_1(t_1, \boldsymbol{\alpha}) & \cdots & \phi_K(t_1, \boldsymbol{\alpha}) \\ \vdots & \ddots & \vdots \\ \phi_1(t_N, \boldsymbol{\alpha}) & \cdots & \phi_K(t_N, \boldsymbol{\alpha}) \end{pmatrix}.$$

Non-linear least squares problems of this type form a special class known as separable non-linear least squares and have been well studied in the numerical analysis community for the past thirty years.

To see how this special structure can be exploited, recall the least squares functional to be minimized is

$$r_1^2(\boldsymbol{\alpha}, \mathbf{a}) = |\mathbf{y} - \mathbf{\Phi}(\boldsymbol{\alpha})\mathbf{a}|^2. \tag{4}$$

Now, suppose we were given *a priori* the value of the non-linear parameters $\boldsymbol{\alpha}$ at the minimum of Eq. (4) which we denote $\hat{\boldsymbol{\alpha}}$. We can easily determine *a posteriori* the linear parameters $\hat{\mathbf{a}}$ by solving the corresponding *linear* least squares problem. The solution is simply

$$\hat{\mathbf{a}} = \mathbf{\Phi}^+(\hat{\boldsymbol{\alpha}})\mathbf{y} \tag{5}$$

where $\mathbf{\Phi}^+(\hat{\boldsymbol{\alpha}})$ is the Moore–Penrose pseudo-inverse of $\mathbf{\Phi}(\hat{\boldsymbol{\alpha}})$ [30]. Substituting Eq. (5) back into Eq. (4) we get a new least squares functional that depends only on $\boldsymbol{\alpha}$

$$r_2^2(\boldsymbol{\alpha}) = \left|\mathbf{y} - \mathbf{\Phi}(\boldsymbol{\alpha})\mathbf{\Phi}^+(\boldsymbol{\alpha})\mathbf{y}\right|^2. \tag{6}$$

$\mathbf{P}(\boldsymbol{\alpha}) \equiv \mathbf{\Phi}(\boldsymbol{\alpha})\mathbf{\Phi}^+(\boldsymbol{\alpha})$ is the orthogonal projector onto the linear space spanned by the column vectors of $\mathbf{\Phi}(\boldsymbol{\alpha})$, so $\mathbf{P}^\perp(\boldsymbol{\alpha}) \equiv \mathbf{1} - \mathbf{P}(\boldsymbol{\alpha})$ is the projector onto the orthogonal complement of the column space of $\mathbf{\Phi}(\boldsymbol{\alpha})$. Hence, we can rewrite Eq. (6) more compactly as

$$r_2^2(\boldsymbol{\alpha}) = \left|\mathbf{P}^\perp(\boldsymbol{\alpha})\mathbf{y}\right|^2.$$

This form makes it easier to see why $r_2^2(\boldsymbol{\alpha})$ is commonly called the *variable projection* (VARPRO) functional. It has been shown [10] that the minima of $r_2(\boldsymbol{\alpha})$ and the corresponding values of **a** from Eq. (5) are equivalent to the minima of $r_1^2(\boldsymbol{\alpha}, \mathbf{a})$.

One complication of the VARPRO method is computing the gradient $\partial \mathbf{r}_2/\partial \boldsymbol{\alpha}$ when the gradients $\partial \phi_k(t_n, \boldsymbol{\alpha})/\partial \boldsymbol{\alpha}$ are known. The solution is presented in some detail in [10] and an excellent FORTRAN implementation [4] is available in the Netlib Repository.

From our review of the NMR spectroscopy literature, it appears that the VARPRO method, and in particular the Netlib implementation, is competitive with the standard least squares method using either the LMDER routine of the MINPACK library or the NL2SOL routines of the PORT library, both also available in the Netlib Repository. In general, the VARPRO functional requires fewer minimizer iterations, but the gradient computation is more expensive. Note that the Levenberg-Marquardt minimizer in [20] performs quite poorly relative to these three and we cannot recommend its use in production code.

Apart from the issue of numerical speed and accuracy of the VARPRO method, we see two additional benefits of this method over the standard method. First, by reducing the dimensionality of the search space by postponing the determination of **â**, this also means that starting estimates for **â** are not needed. For LQCD, this is a great benefit, since good guesses for $\boldsymbol{\alpha}$ are easily obtained from the black box methods of Sec. 3. Second, when the incorporation of Bayesian prior knowledge is desired, for LQCD it seems easier to develop reasonable priors for the energies E_k than the amplitudes A_k. When using the VARPRO method, only priors for the energies are needed. Of course, if reliable priors for the amplitudes are available, one should instead use the standard method. Finally, data covariance can easily be incorporated in the usual way

$$r_2^2(\boldsymbol{\alpha}) = \left[\mathbf{P}^\perp(\boldsymbol{\alpha})\mathbf{y}\right]^T \mathbf{C}^{-1}(\mathbf{y}) \left[\mathbf{P}^\perp(\boldsymbol{\alpha})\mathbf{y}\right].$$

3 Black Box Methods

3.1 Effective Masses

The best example of a black box method widely used in LQCD is the method of effective masses. Let's consider the problem of Eq. (3) for the case $N=2$, $K=1$

$$\begin{pmatrix} y_n \\ y_{n+1} \end{pmatrix} = \begin{pmatrix} \alpha_1^n \\ \alpha_1^{n+1} \end{pmatrix}(a_1) \quad \Rightarrow \quad \alpha_1 = \frac{y_{n+1}}{y_n}, \quad a_1 = \frac{y_n}{\alpha_1^n}$$

As expected, the problem is exactly determined, so there is an unique zero residual solution. For the model function of Eq. (1) the effective mass is $m_{\text{eff}} = -\log(\alpha_1)$. Note that the non-linear parameter α_1 is determined first from the data and then the linear parameter a_1 can be determined. This is an indication of the separability of the least squares problem discussed in Sec. 2.

As we are unaware of its presentation elsewhere, here is the two-state effective mass solution. We start from Eq. (3) for $N=4$, $K=2$

$$\begin{pmatrix} y_n \\ y_{n+1} \\ y_{n+2} \\ y_{n+3} \end{pmatrix} = \begin{pmatrix} 1 & 1 \\ \alpha_1 & \alpha_2 \\ \alpha_1^2 & \alpha_2^2 \\ \alpha_1^3 & \alpha_2^3 \end{pmatrix} \begin{pmatrix} a_1 \alpha_1^n \\ a_2 \alpha_2^n \end{pmatrix}. \qquad (7)$$

If we compute three quantities from the data

$$A = y_{n+1}^2 - y_n y_{n+2}$$
$$B = y_n y_{n+3} - y_{n+1} y_{n+2}$$
$$C = y_{n+2}^2 - y_{n+1} y_{n+3}$$

then the two solutions for the non-linear parameters α_k come from the familiar quadratic equation

$$\alpha_{1,2} = \frac{-B \pm \sqrt{B^2 - 4AC}}{2A}. \qquad (8)$$

As before, the linear parameters $a_{1,2}$ can also be determined once the non-linear parameters are known

$$a_k \alpha_k^n = \frac{1}{2} \left[y_n \pm \frac{\sqrt{(B^2 - 4AC)[4A^3 + (B^2 - 4AC)y_n^2]}}{B^2 - 4AC} \right]$$

where some care must be taken to properly match solutions.

In general, when $N=2K$ there should always be such a unique zero residual solution. From inspection of Eq. (7) the $N=4$, $K=2$ problem is a set of 4 coupled cubic equations. Unfortunately, due to Abel's Impossibility Theorem [2], we should expect that general algebraic solutions are only possible for $N \leq 5$. Yet, the rather surprising result of Eq.(8) is that after properly separating the non-linear parameters α_k, the $N=4$, $K=2$ problem is of quadratic order. Thus, we suspect that it is also possible to find algebraic solutions to the three-state and four-state effective mass problems when properly reduced to cubic and quartic equations after separation of variables.

3.2 Black Box I: Linear Prediction

In order to compute solutions of Eq. (3) when the system is over-determined ($N > 2K$) or when an algebraic solution is not available, we consider the first black box method called *linear prediction*. We form a K-th order polynomial with the α_k as roots

$$p(\alpha) = \prod_{k=1}^{K} (\alpha - \alpha_k) = \sum_{i=0}^{K} p_i \alpha^{K-i} \quad (p_0 = 1). \qquad (9)$$

Since $p(\alpha_k) = 0$ the following is true

$$\alpha_k^m = -\sum_{i=1}^{K} p_i \, \alpha_k^{m-i}, \quad m \geq K. \qquad (10)$$

When Eq. (10) is substituted in Eq. (3) we find the following relation

$$y_m = -\sum_{k=1}^{K} p_k \, y_{m-k}, \quad m \geq K. \qquad (11)$$

Because Eq. (11) enables us to "predict" the data y_m at larger times in terms of the data $y_{m-K}, \cdots, y_{m-1}$ at earlier times, the p_k are commonly called *forward linear prediction coefficients*.

Using Eq. (11) we can construct the linear system $\mathbf{h}_{\text{lp}} = -\mathbf{H}_{\text{lp}}\mathbf{p}$

$$\begin{pmatrix} y_M \\ y_{M+1} \\ \vdots \\ y_{M-1} \end{pmatrix} = - \begin{pmatrix} y_0 & \cdots & y_{M-1} \\ y_1 & \cdots & y_M \\ \vdots & \ddots & \vdots \\ y_{N-M-1} & \cdots & y_{N-2} \end{pmatrix} \begin{pmatrix} p_M \\ p_{M-1} \\ \vdots \\ p_1 \end{pmatrix}, \quad N \geq 2M. \qquad (12)$$

In numerical analysis, this is known as a Hankel system and the matrix $\mathbf{H}_{\text{lp}}$ is a Hankel matrix. After solving Eq. (12) for $\mathbf{p}$, the roots of the polynomial of Eq. (9) are computed to determine the parameters α_k. The a_k parameters can subsequently be determined from Eq. (5).

In the presence of noisy data, the equality in Eq. (12) is only approximate, even for the case $N = 2M$, so some minimization method like least squares must be used. This doesn't mean that the parameter estimates from linear prediction agree with the parameter estimates from the least squares methods of Sec. 2. Gauss proved [8] that the least squares estimates of fit parameters for linear problems have the smallest possible variance. In this sense, least squares estimates are considered *optimal* although we know of no proof that this holds for non-linear problems. Since linear prediction estimates may not agree with least squares, they are considered *sub-optimal* even though there is no proof that the variance is larger, due to non-linearity.

A popular method for solving Eq. (12) is the LPSVD algorithm [14]. In this method, we construct $\mathbf{H}_{\text{lp}}$ for M as large as possible, even if we are only interested in estimating $K < M$ parameters. After computing the SVD of $\mathbf{H}_{\text{lp}}$, we construct a rank K approximation $\mathbf{H}_{\text{lp}K}$

$$\mathbf{H}_{\text{lp}} = \mathbf{U}\mathbf{\Sigma}\mathbf{V}^\dagger = (\mathbf{U}_K \mathbf{U}_2) \begin{pmatrix} \mathbf{\Sigma}_K & \\ & \mathbf{\Sigma}_2 \end{pmatrix} (\mathbf{V}_K \mathbf{V}_2)^\dagger, \quad \mathbf{H}_{\text{lp}K} = \mathbf{U}_K \mathbf{\Sigma}_K \mathbf{V}_K^\dagger.$$

$\mathbf{\Sigma}_K$ contains the K largest singular values. By zeroing $\mathbf{\Sigma}_2$ to reduce the rank of $\mathbf{H}_{\text{lp}}$, much of the statistical noise is eliminated from the problem. From the Eckart–Young–Mirsky theorem [6, 17], this rank K approximation is the nearest possible under either the Frobenius norm or matrix 2-norm. Then, after solving $\mathbf{h}_{\text{lp}} = -\mathbf{H}_{\text{lp}K}\mathbf{p}$ for the p_m coefficients, the M roots of the polynomial in Eq. (9) are computed using a root-finding algorithm. Since the rank of $\mathbf{H}_{\text{lp}}$ was reduced to K, only K roots are valid parameter estimates. Typically, the K largest magnitude roots are chosen.

Our experience with this algorithm is that the largest magnitude roots often have unphysical values if K is set larger than a reasonable number given the statistical precision of the data. There are also several issues which may be of some concern. First, we found that root-finding algorithms all come with caveats about stability and susceptibility to round-off error and must be treated with some care. Also, since statistical noise is present on both sides of Eq. (12), the rank-reduced least squares solution is probably not appropriate and one should probably use an errors-in-variables (EIV) approach like total least squares (TLS), which we will describe in Sec. 3.3. We have found that the TLS variant of LPSVD, called LPTLS [21], gives better parameter estimates than LPSVD.

3.3 Total Least Squares

In the standard linear least squares problem $\mathbf{Ax} \approx \mathbf{b}$

$$\underset{\mathbf{x} \in \mathbb{R}^K}{\text{minimize}} \|\mathbf{Ax} - \mathbf{b}\|_2, \quad \mathbf{A} \in \mathbb{R}^{N \times K}, \, \mathbf{b} \in \mathbb{R}^N, \, N \geq K \tag{13}$$

an important assumption for the solution, *i.e.* Eq. (5), to be considered optimal is that the only errors are in the data vector $\mathbf{b}$ and further that those errors are *i.i.d.* (independent and identically distributed). When errors also occur in $\mathbf{A}$, as in Eq. (12), then a new approach, often called errors-in-variables (EIV), is required to restore optimality. Note that the errors in $\mathbf{A}$ that cause the loss of optimality need not be purely statistical: numerical round-off errors or choosing to fit a model function which differs from the "true" model function are potential sources of error which could cause loss of optimality.

To understand the total least squares (TLS) solution to the EIV problem, consider the case when a zero residual solution to Eq. (13) exists. Then, if we add $\mathbf{b}$ as a column of $\mathbf{A}$, written $[\mathbf{Ab}]$, it cannot have greater column rank than $\mathbf{A}$ because $\mathbf{b} \in \text{Ran}(\mathbf{A})$. If we compute the SVD of $[\mathbf{Ab}]$ we will find that the singular value $\sigma_{K+1} = 0$. When the solution of Eq. (13) has non-zero residual, we may find the singular value σ_{K+1} of $[\mathbf{Ab}]$ to be non-zero as well. But, we can construct the nearest rank $R \leq K$ approximation to $[\mathbf{Ab}]$ (in the sense of the Eckart–Young–Mirsky theorem) and this gives us the TLS solution. The TLS solution was computed in [11, 9], although the name was not coined until [12]. A comprehensive review [28] of the subject is available.

Finally, TLS is very sensitive to the distribution of errors in $[\mathbf{Ab}]$. If the errors are not known to be i.i.d. then it is crucial to scale the matrix before using the TLS algorithm. If the data are uncorrelated, then a method known as *"equilibrium"* scaling [3] is sufficient. If the data are correlated, then Cholesky factors of the covariance matrix must be used. In this case, it is better to use either the generalized TLS algorithm (GTLS) [24, 25] or the restricted TLS algorithm (RTLS) [29] which are more robust when the covariance matrix is ill-conditioned. Implementations of various TLS algorithms are available in the Netlib Repository [23].

3.4 Black Box II: State Space Methods

The name for these methods is derived from state-space theory in the control and identification literature [13]. The basic approach is to compute the non-linear parameters α_k of Eq. (3) without needing to compute the roots of a polynomial, as in Sec. 3.2. From Eq. (12), we start by noting that $\mathbf{H}_s = [\mathbf{H}_{lp} \mathbf{h}_{lp}]$ is also a Hankel matrix

$$\mathbf{H}_s = \begin{pmatrix} y_0 & \cdots & y_{M-1} & y_M \\ \vdots & \ddots & \vdots & \vdots \\ y_{N-M-1} & \cdots & y_{N-2} & y_{N-1} \end{pmatrix} \quad M \geq K, \, N - M > K.$$

A Vandermonde decomposition exists for this matrix

$$\mathbf{SAT}^T = \begin{pmatrix} 1 & \cdots & 1 \\ \alpha_1 & \cdots & \alpha_K \\ \vdots & \ddots & \vdots \\ \alpha_1^{N-M-1} & \cdots & \alpha_K^{N-M-1} \end{pmatrix} \begin{pmatrix} a_1 & & \\ & \ddots & \\ & & a_K \end{pmatrix} \begin{pmatrix} 1 & \cdots & 1 \\ \alpha_1 & \cdots & \alpha_K \\ \vdots & \ddots & \vdots \\ \alpha_1^M & \cdots & \alpha_K^M \end{pmatrix}^T. \tag{14}$$

in terms of the linear (a_k) and non-linear (α_k) parameters of Eq. (3). If we could compute this decomposition directly, then the problem would be solved. Alas, no such algorithm is currently known.

An indirect method exists to compute this decomposition called Hankel SVD (HSVD). We will consider here a TLS variant called HTLS [27]. First, we note the shift invariance property of $\mathbf{S}$ (and similarly for $\mathbf{T}$)

$$\mathbf{S}^\uparrow \mathcal{A} = \mathbf{S}_\downarrow, \qquad \mathcal{A} = \mathbf{diag}(\alpha_1, \cdots, \alpha_K).$$

Next, we note that if such a decomposition is possible, then $\mathbf{S}$, $\mathcal{A}$ and $\mathbf{T}$ are all of rank K by inspection, so $\mathbf{H}_s$ is at least of rank K, as well. So, using SVD we construct the nearest rank K approximation to $\mathbf{H}_{sK}$

$$\mathbf{H}_{sK} = (\mathbf{U}_K \mathbf{U}_2) \begin{pmatrix} \mathbf{\Sigma}_K & \\ & 0 \end{pmatrix} (\mathbf{V}_K \mathbf{V}_2)^\dagger = \mathbf{U}_K \mathbf{\Sigma}_K \mathbf{V}_K^\dagger. \qquad (15)$$

By comparing the decompositions of Eq. (14) and Eq. (15) we can see

$$\mathrm{Span}(\mathbf{S}) = \mathrm{Span}(\mathbf{U}_K) \implies \mathbf{U}_K = \mathbf{SQ} \implies \mathbf{U}_K^\uparrow = \mathbf{U}_{K\downarrow} \mathbf{Q}^{-1} \mathcal{A} \mathbf{Q}.$$

So, computing the TLS solution of $\left[\mathbf{U}_K^\uparrow \mathbf{U}_{K\downarrow}\right]$ will give us $\mathbf{Q}^{-1} \mathcal{A} \mathbf{Q}$, which we can then diagonalize using an eigenvalue solver to get our estimates of α_k.

In our experience with these black box methods, the HTLS algorithm seems to be the most robust. However, we would like to emphasize two points.

First, the estimates of α_k from HTLS are considered sub-optimal because $\mathbf{H}_{sK}$ in Eq. (15) is only approximately, but not *exactly*, a Hankel matrix because the SVD does not enforce the Hankel structure throughout. A similar problem occurs while constructing the TLS solution of $\left[\mathbf{U}_K^\uparrow \mathbf{U}_{K\downarrow}\right]$. *Structured* TLS algorithms (STLS) exist which can construct $\mathbf{H}_{sK}$ while preserving the Hankel structure (see [26] for references) and hence restoring the optimality of the estimates. While we have not yet tried these STLS algorithms, we note that all of them involve iterative procedures to restore the structure. Thus, under the *"no free lunch"* theorem, we suspect that the price of restoring optimality is roughly equivalent to performing the (optimal) non-linear least squares minimizations described in Sec. 2.

Our second observation is that LQCD data is always correlated, so that a GTLS or RTLS algorithm is needed to compute the TLS solution of $\left[\mathbf{U}_K^\uparrow \mathbf{U}_{K\downarrow}\right]$. But, covariance estimates of $\mathbf{U}_K$ are not readily computed from the data covariance matrix because of the required SVD. Thus, a jackknife or bootstrap resampling method is required to estimate $\mathbf{cov}(\mathbf{U}_K)$. Since we expect to use a resampling method to estimate the covariance of the α_k, this means that there is an inner and outer resampling loop so the method can easily become computationally expensive if the number of data samples becomes large. In this case, blocking the data is recommended.

4 Conclusions

We have found that reviewing the literature of other fields where data analysis of exponentially damped time series is also prevalent to be quite fruitful. Our review has discovered several mature analysis methods which are virtually unknown (or unmentioned) in the Lattice QCD literature. We have performed several tests of all the methods discussed on fake data and on some actual LQCD data are encouraged by the results. So, we are incorporating these techniques into our production versions of analysis programs and expect to report results soon.

Finally, we would like to acknowledge that we have found Leentje Vanhamme's Ph.D. Thesis [26] an extremely useful guide to the literature of the NMR spectroscopy community. We would encourage anyone interested in learning more to start there. An electronic copy is currently available online.

This work was supported in part by DOE contract DE-AC05-84ER40150 under which the Southeastern Universities Research Association (SURA) operates the Thomas Jefferson National Accelerator Facility.

References

1. Chris Allton, Danielle Blythe, and Jonathan Clowser. Spectral functions, maximum entropy method and unconventional methods in lattice field theory. *Nucl. Phys. Proc. Suppl.*, 109A:192–196, 2002.
2. Niels H. Abel. Beweis der Unmöglichkeit, algebraische Gleichungen von höheren Graden als dem vierten allgemein aufzulösen. *J. reine angew. Math.*, 1:65, 1826. See http://mathworld.wolfram.com/AbelsImpossibilityTheorem.html for more references.
3. F. L. Bauer. Optimally scaled matrices. *Numer. Math.*, 5:73–87, 1963.
4. John Bolstad. varpro.f. Available at: http://www.netlib.org/opt/index.html, January 1977.
5. T. Draper et al. An algorithm for obtaining reliable priors for constrained- curve fits. 2003.
6. Carl H. Eckart and Gale Young. The approximation of one matrix by another of lower rank. *Psychometrika*, 1:211–218, 1936.
7. H. Rudolf Fiebig. Spectral density analysis of time correlation functions in lattice QCD using the maximum entropy method. *Phys. Rev.*, D65:094512, 2002.
8. Carl F. Gauss. Theoria combinationis observationum erroribus minimis obnoxiae. *Comment. Soc. Reg. Sci. Gotten. Recent.*, 5:33, 1823.
9. Gene H. Golub. Some modified matrix eigenvalue problems. *SIAM Rev.*, 15:318–334, 1973.
10. Gene H. Golub and Victor Pereyra. The differentiation of pseudoinverses and nonlinear least squares problems whose variables separate. *SIAM J. Numer. Anal.*, 10:413–432, 1973.
11. Gene H. Golub and Christian H. Reinsch. Singular value decomposition and least squares solutions. *Numer. Math.*, 14:403–420, 1970.
12. Gene H. Golub and Charles F. Van Loan. An analysis of the total least squares problem. *SIAM J. Numer. Anal.*, 17:883–893, 1980.
13. S. Y. Kung, K. S. Arun, and D. V. Bhaskar Rao. State-space and singular value decomposition-based approximation methods for the harmonic retrieval problem. *J. Opt. Soc. Am.*, 73:1799–1811, 1983.
14. Ramdas Kumeresan and Donald W. Tufts. Estimating the parameters of exponentially damped sinusoids and pole-zero modeling in noise. *IEEE Trans. Acoust. Speech Signal Proc.*, 30:833–840, 1982.

15. G. P. Lepage et al. Constrained curve fitting. *Nucl. Phys. Proc. Suppl.*, 106:12–20, 2002.
16. Chris Michael. Fitting correlated data. *Phys. Rev.*, D49:2616–2619, 1994.
17. Leon Mirsky. Symmetric gauge functions and unitarily invariant norms. *Quart. J. Math. Oxford Ser.*, 11:50–59, 1960.
18. Chris Michael and A. McKerrell. Fitting correlated hadron mass spectrum data. *Phys. Rev.*, D51:3745–3750, 1995.
19. Y. Nakahara, M. Asakawa, and T. Hatsuda. Hadronic spectral functions in lattice QCD. *Phys. Rev.*, D60:091503, 1999.
20. William H. Press, Brian P. Flannery, Saul A. Teukolsky, and William T. Vetterling. *Numerical Recipes: The Art of Scientific Computing*. Cambridge University Press, Cambridge (UK) and New York, second edition, 1992.
21. C. F. Tirendi and J. F. Martin. Quantitative analysis of NMR spectra by linear prediction and total least squares. *J. Magn. Reson.*, 85:162–169, 1989.
22. D. Toussaint. In T. DeGrand and D. Toussaint, editors, *From Actions to Answers*, page 121, Singapore, 1990. World Scientific. Proceedings of Theoretical Advanced Study Institute in Elementary Particle Physics, Boulder, USA, June 5-30, 1989.
23. Sabine Van Huffel. Available at: http://www.netlib.org/vanhuffel/, 1988.
24. Sabine Van Huffel. The generalized total least squares problem: formulation, algorithm and properties. In Gene H. Golub and P. Van Dooren, editors, *Numerical Linear Algebra, Digital Signal Processing and Parallel Algorithms*, volume 70 of *NATO ASI Series F: Computer and Systems Sciences*, pages 651–660, Berlin, 1990. Springer-Verlag. Proceedings of NATO ASI, Leuven, Belgium, August 1988.
25. Sabine Van Huffel. Reliable and efficient techniques based on total least squares for computing consistent estimators in models with errors in the variables. In J. G. McWhirter, editor, *Mathematics in Signal Processing II*, pages 593–603, Oxford, 1990. Clarendon Press. Proceedings of IMA conference, December 1988.
26. Leentje Vanhamme. *Advanced time-domain methods for nuclear magnetic resonance spectroscopy data analysis*. PhD thesis, Katholieke Universiteit Leuven, Belgium, November 1999. ftp://ftp.esat.kuleuven.ac.be/pub/sista/vanhamme/reports/phd.ps.gz.
27. Sabine Van Huffel, H. Chen, C. Decanniere, and P. Van Hecke. Algorithm for time-domain NMR data fitting based on total least squares. *J. Magn. Reson., Ser. A*, 110:228–237, 1994.
28. Sabine Van Huffel and Joos Vandewalle. *The Total Least Squares Problem: Computational Aspects and Analysis*, volume 9 of *Frontiers in Applied Mathematics*. SIAM, Philadelphia, 1992.
29. Sabine Van Huffel and H. Zha. The restricted total least squares problem: formulation, algorithm and properties. *SIAM J. Matrix Anal. Appl.*, 12:292–309, 1991.
30. Eric W. Weisstein. Moore-Penrose matrix inverse. http://mathworld.wolfram.com/Moore-PenroseMatrixInverse.html, 2004. From MathWorld – A Wolfram Web Resource.

Numerical Methods for the QCD Overlap Operator: II. Optimal Krylov Subspace Methods

Guido Arnold[1], Nigel Cundy[1], Jasper van den Eshof[2], Andreas Frommer[3], Stefan Krieg[1], Thomas Lippert[1], and Katrin Schäfer[3]

[1] Department of Physics, University of Wuppertal, Germany
{arnold,cundy,krieg,lippert}@theorie.physik.uni-wuppertal.de
[2] Department of Mathematics, University of Düsseldorf, Germany
eshof@am.uni-duesseldorf.de
[3] Department of Mathematics, University of Wuppertal, Germany
{frommer,schaefer}@math.uni-wuppertal.de

Summary. We investigate optimal choices for the (outer) iteration method to use when solving linear systems with Neuberger's overlap operator in QCD. Different formulations for this operator give rise to different iterative solvers, which are optimal for the respective formulation. We compare these methods in theory and practice to find the overall optimal one. For the first time, we apply the so-called SUMR method of Jagels and Reichel to the shifted unitary version of Neuberger's operator, and show that this method is in a sense the optimal choice for propagator computations. When solving the "squared" equations in a dynamical simulation with two degenerate flavours, it turns out that the CG method should be used.

1 Introduction

Recently, lattice formulations of QCD respecting chiral symmetry have attracted a lot of attention. A particular promising such formulation, the so-called overlap fermions, has been proposed in [9]. From the computational point of view, we have to solve linear systems involving the sign function $\text{sign}(Q)$ of the (hermitian) Wilson fermion matrix Q. These computations are very costly, and it is of vital importance to devise efficient numerical schemes.

A direct computation of $\text{sign}(Q)$ is not feasible, since Q is large and sparse, whereas $\text{sign}(Q)$ would be full. Therefore, numerical algorithms have to follow an inner-outer paradigm: One performs an outer Krylov subspace method where each iteration requires the computation of a matrix-vector product involving $\text{sign}(Q)$. Each such product is computed through another, inner iteration using matrix-vector multiplications with Q. In an earlier paper [12] we investigated methods for the inner iteration and established the Zolotarev rational approximation together with the multishift CG method [5] as the method of choice.

In the present paper we investigate optimal methods for the outer iteration. We consider two situations: the case of a propagator computation and the case of a pseudofermion computation within a dynamical hybrid Monte Carlo simulation where one has to solve the "squared" system. As we will see, the optimal method for the case of a propagator computation is a not so well-known method due to Jagels and Reichel [8], whereas in the case of the squared system it will be best to apply classical CG on that squared system rather than using a two-pass approach.

This paper is organized as follows: We first introduce our notation in Section 2. We then discuss different equivalent formulations of the Neuberger overlap operator and establish useful relations between the eigenpairs of these different formulations (Section 3). We discuss optimal Krylov subspace methods for the various systems in Section 4 and give some theoretical results on their convergence speed based on the spectral information from Section 3. In Section 5 we compare the convergence speeds both, theoretically and in practical experiments. Our conclusions will be summarized in Section 6.

2 Notation

The Wilson-Dirac fermion operator,

$$M = I - \kappa D_W,$$

represents a nearest neighbour coupling on a four-dimensional space-time lattice, where the "hopping term" D_W is a non-normal sparse matrix, see (17) in the appendix. The coupling parameter κ is a real number which is related to the bare quark mass.

The massless overlap operator is defined as

$$D_0 = I + M \cdot (M^\dagger M)^{-\frac{1}{2}},$$

where $M^\dagger$ denotes the conjugate transpose of M. For the massive overlap operator, for notational convenience we use a mass parameter $\rho > 1$ such that this operator is given as

$$D = \rho I + M \cdot (M^\dagger M)^{-\frac{1}{2}}. \tag{1}$$

In the appendix, we explain that this form is just a scaled version of Neuberger's original choice, and we relate ρ to the quark mass, see (21) and (22).

Replacing M in (1) by its hermitian form Q, see (18), the overlap operator can equivalently be written as

$$D = \rho I + \gamma_5 \operatorname{sign}(Q) = \gamma_5 \cdot (\rho \gamma_5 + \operatorname{sign}(Q)),$$

with γ_5 being defined in (19) and $\text{sign}(Q)$ being the standard matrix sign function. Note that $\rho\gamma_5 + \text{sign}(Q)$ is hermitian and indefinite, whereas $\gamma_5 \text{sign}(Q)$ is unitary.

To reflect these facts in our notation, we define:

$$D_u = \rho I + \gamma_5 \text{sign}(Q), \qquad D_h = \rho\gamma_5 + \text{sign}(Q),$$

where $D_u = \gamma_5 D_h$. Both these operators are normal, i.e. they commute with their adjoints.

3 Formulations and Their Spectral Properties

3.1 Propagator Computations

When computing quark propagators, the systems to solve are of the form

$$D_u x = (\rho I + \gamma_5 \text{sign}(Q))x = b. \tag{2}$$

Multiplying this shifted unitary form by γ_5, we obtain its *hermitian indefinite form* as

$$D_h x = (\rho\gamma_5 + \text{sign}(Q))x = \gamma_5 b. \tag{3}$$

The two operators D_u and D_h are intimately related. They are both normal, and as a consequence the eigenvalues (and the weight with which the corresponding eigenvectors appear in the initial residual) solely govern the convergence behaviour of an optimal Krylov subspace method used to solve the respective equation.

Very interestingly, the eigenvalues of the different operators can be explicitly related to each other. This allows a quite detailed discussion of the convergence properties of adequate Krylov subspace solvers. To see this, let us introduce an auxiliary decomposition for $\text{sign}(Q)$: Using the chiral representation, the matrix γ_5 on the whole lattice can be represented as a 2×2 block diagonal matrix

$$\gamma_5 = \begin{pmatrix} I & 0 \\ 0 & -I \end{pmatrix}, \tag{4}$$

where both diagonal blocks I and $-I$ are of the same size. Partitioning $\text{sign}(Q)$ correspondingly gives

$$\text{sign}(Q) = \begin{pmatrix} S_{11} & S_{12} \\ S_{12}^\dagger & S_{22} \end{pmatrix}. \tag{5}$$

In Lemma 1 below we give a convenient decomposition for this matrix that is closely related to the so-called *CS decomposition* (see [6, Theorem 2.6.3] and the references therein), an important tool in matrix analysis. Actually, the lemma may be regarded as a variant of the CS decomposition for hermitian matrices where the decomposition here can be achieved using a similarity transform. The proof follows the same lines as the proof for the existence of the CS decomposition given in [11].

Lemma 1. *There exists a unitary matrix X such that*

$$\text{sign}(Q) = X \begin{pmatrix} \Phi & \Sigma \\ \Sigma & \Psi \end{pmatrix} X^\dagger, \quad \text{with } X = \begin{pmatrix} X_1 & 0 \\ 0 & X_2 \end{pmatrix}.$$

The matrices Φ, Ψ and Σ are real and diagonal with diagonal elements ϕ_j, ψ_j and $\sigma_j \geq 0$, respectively. Furthermore, $\phi_j^2 + \sigma_j^2 = \psi_j^2 + \sigma_j^2 = 1$ and

$$\phi_j = -\psi_j \quad \text{if} \quad \sigma_j > 0 \tag{6}$$
$$\phi_j, \psi_j \in \{-1, +1\} \quad \text{if} \quad \sigma_j = 0. \tag{7}$$

Note that in the case of (6) we know that ϕ_j and ψ_j have opposite signs, whereas in case (7) we might have $\phi_j = \psi_j = 1$ or $\phi_j = \psi_j = -1$. The key point for X in Lemma 1 is the fact that $\gamma_5 X = X \gamma_5$. This allows us to relate the eigenvalues and -vectors of the different formulations for the overlap operator via ϕ_j, ψ_j and σ_j. In this manner we give results complementary to Edwards et al. [2], where relations between the eigenvalues (and partly the -vectors) of the different formulations for the overlap operator were given without connecting them to sign(Q) via ϕ_j, ψ_j and σ_j. The following lemma gives expressions for the eigenvectors and -values of the shifted unitary operator D_u.

Lemma 2. *With the notation from Lemma 1, let x_j^1 and x_j^2 be the j-th column of X_1 and X_2, respectively. Then $\text{spec}(D_u) = \{\lambda_{j,\pm}^u\}$ with*

$$\lambda_{j,\pm}^u = \rho + \phi_j \pm i\sqrt{1 - \phi_j^2} \quad \text{if} \quad \sigma_j \neq 0$$
$$\lambda_{j,+}^u = \rho + \phi_j, \; \lambda_{j,-}^u = \rho - \psi_j \quad \text{if} \quad \sigma_j = 0.$$

The corresponding eigenvectors are

$$z_{j,\pm}^u = \begin{pmatrix} \mp i x_j^1 \\ x_j^2 \end{pmatrix} \quad \text{if} \quad \sigma_j \neq 0$$
$$z_{j,+}^u = \begin{pmatrix} x_j^1 \\ 0 \end{pmatrix}, \; z_{j,-}^u = \begin{pmatrix} 0 \\ x_j^2 \end{pmatrix} \quad \text{if} \quad \sigma_j = 0.$$

Proof. With the decomposition for sign(Q) in Lemma 1, we find, using $X^\dagger \gamma_5 = \gamma_5 X^\dagger$,

$$X^\dagger \gamma_5 \text{sign}(Q) X = \gamma_5 X^\dagger \text{sign}(Q) X = \begin{pmatrix} \Phi & \Sigma \\ -\Sigma & -\Psi \end{pmatrix}.$$

Since the actions of X and $X^\dagger$ represent a similarity transform, we can easily derive the eigenvalues and eigenvectors of $\gamma_5 \text{sign}(Q)$ from the matrix on the right. This can be accomplished by noticing that this matrix can be permuted to have a block diagonal form with 2×2 blocks on the diagonal, so that the problem reduces to the straightforward computation of the eigenvalues and eigenvectors of 2×2 matrices. This concludes the proof.

As a side remark, we mention that Edwards et al. [2] observed that the eigenvalues of D_u can be efficiently computed by exploiting the fact that most of the eigenvalues come in complex conjugate pairs. Indeed, using Lemma 1 and the fact that $\gamma_5 X = X\gamma_5$ we see that we only have to compute the eigenvalues of the hermitian matrix S_{11} (and to check for 1 or -1 eigenvalues in S_{22}).

With the same technique as in Lemma 2 we can also find expressions for the eigenvalues and eigenvectors of the hermitian indefinite formulation.

Lemma 3. *With the same notation as in Lemma 2, we have that* $\operatorname{spec}(D_h) = \{\lambda^h_{j,\pm}\}$ *with*

$$\lambda^h_{j,\pm} = \pm\sqrt{1 + 2\phi_j\rho + \rho^2} \quad \text{if} \quad \sigma_j \neq 0$$
$$\lambda^h_{j,+} = \rho + \phi_j, \; \lambda^h_{j,-} = -\rho + \psi_j \quad \text{if} \quad \sigma_j = 0.$$

The corresponding eigenvectors are

$$z^h_{j,\pm} = \begin{pmatrix} (\phi_j\rho + \lambda^h_{j,\pm})x^1_j \\ \sqrt{1-\phi_j^2}\, x^2_j \end{pmatrix} \quad \text{if} \quad \sigma_j \neq 0,$$

$$z^h_{j,+} = \begin{pmatrix} x^1_j \\ 0 \end{pmatrix}, \; z^h_{j,-} = \begin{pmatrix} 0 \\ x^2_j \end{pmatrix} \quad \text{if} \quad \sigma_j = 0.$$

As an illustration to the results of this section, we performed numerical computations for two sample configurations. Both are on a 4^4 lattice with $\beta = 6.0$ (configuration A) and $\beta = 5.0$ (configuration B) respectively. [4] Figure 1 shows plots of the eigenvalues of M for these configurations. We used $\kappa = 0.2129$ for configuration A, $\kappa = 0.2809$ for configuration B.

Figure 2 gives plots of the eigenvalues of D_u and D_h for our example configurations with $\rho = 1.01$. It illustrates some interesting consequences of Lemma 2 and Lemma 3: With $C(\rho, 1)$ denoting the circle in the complex plane with radius 1 centered at ρ, we have

$$\operatorname{spec}(D_u) \subseteq C(\rho, 1),$$

and, moreover, $\operatorname{spec}(D_u)$ is symmetric w.r.t. the real axis. On the other hand, $\operatorname{spec}(D_h)$ is "almost symmetric" w.r.t. the origin, the only exceptions corresponding to the case $\sigma_j = 0$, where an eigenvalue of the form $\rho + \phi_j$ with $\phi_j \in \{\pm 1\}$ not necessarily matches $-\rho + \psi_j$ with $\psi_j \in \{\pm 1\}$, since we do not necessarily have $\psi_j = -\phi_j$. Moreover,

$$\operatorname{spec}(D_h) \subseteq [-(\rho+1), -(\rho-1)] \cup [(\rho-1), (\rho+1)]. \tag{8}$$

---

[4]The hopping matrices for both configurations are available at www.math.uni-wuppertal.de/org/SciComp/preprints.html as well as `matlab` code for all methods presented here. The configurations are also available at Matrix Market as `conf6.0_0014x4.2000.mtx` and `conf5.0_0014x4.2600.mtx` respectively.

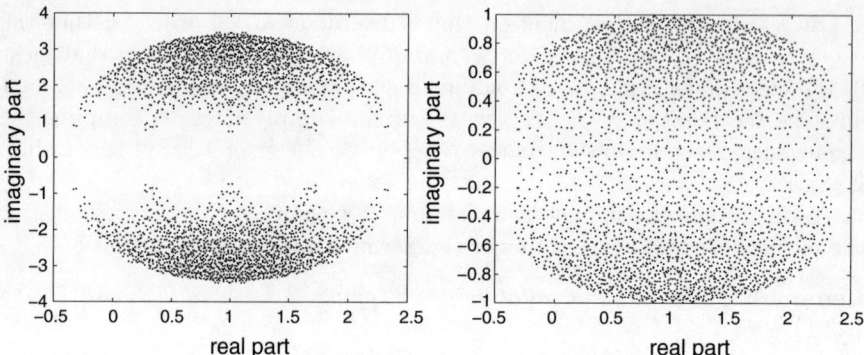

Fig. 1. Eigenvalues for the Wilson fermion matrix M for configurations A (left) and B (right)

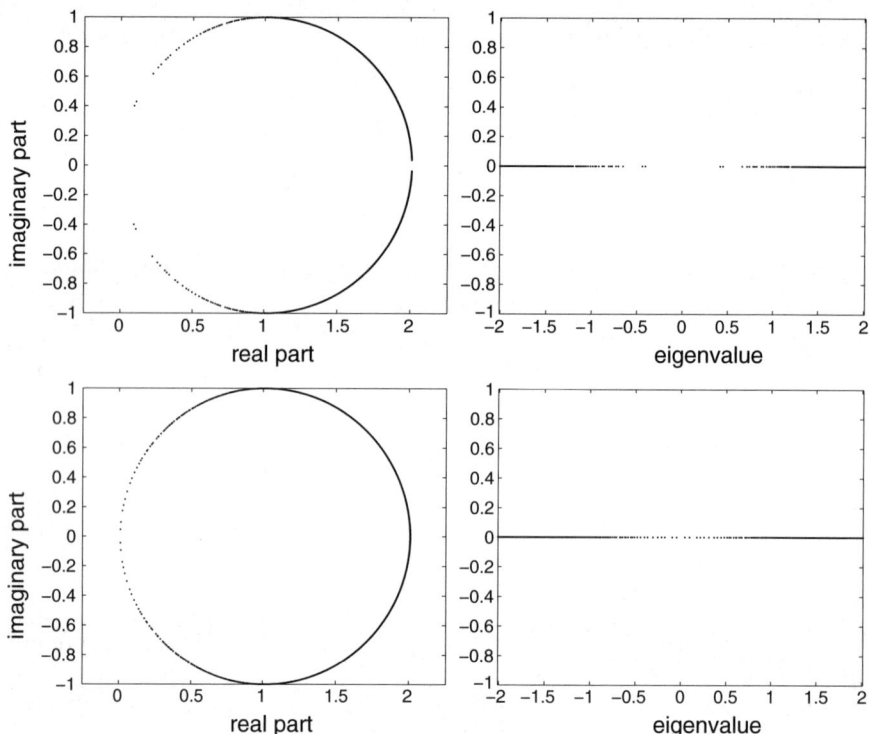

Fig. 2. Eigenvalues of D_u (left column) and D_h (right column) all for $\rho = 1.01$. The upper plots are for configuration A, the lower for configuration B.

Finally, let us note that we have some σ_j equal to zero as soon as 0 is an eigenvalue of the massless operator $\gamma_5 + \text{sign}(Q)$, i.e., as soon as the configuration has non-trivial topology.

3.2 Dynamical Simulations

In a simulation with dynamical fermions, the costly computational task is the inclusion of the fermionic part of the action into the "force" evolving the gauge fields. This requires to solve the "squared" system

$$D_u^\dagger D_u x = b \iff D_h^2 x = b.$$

We will denote D_n the respective operator, i.e.,

$$D_n = D_h^2 = D_u^\dagger D_u.$$

If we plug in the definition of D_u, we find that

$$D_n y = D_u^\dagger D_u y = \left((\rho^2 + 1)I + \rho(\gamma_5 \text{sign}(Q) + \text{sign}(Q)\gamma_5)\right) y = b, \quad (9)$$

from which we get the interesting relation

$$D_u D_u^\dagger = \gamma_5 D_u^\dagger D_u \gamma_5 = D_u^\dagger D_u. \quad (10)$$

As is well known, (9) becomes block diagonal, since, using (4) and (5), we have

$$D_n = \begin{pmatrix} (\rho^2 + 1)I + 2\rho S_{11} & 0 \\ 0 & (\rho^2 + 1)I - 2\rho S_{22} \end{pmatrix} y = b. \quad (11)$$

In practical simulations, the decoupled structure of (11) can be exploited by running two separate CG processes simultaneously, one for each (half-size) block. Since each of these CG processes only has to accommodate a part of the spectrum of D_n, convergence will be faster. An alternative usage of the block structure is to compute the action of the matrix sign function on a two-dimensional subspace corresponding to the two blocks and to use a block type Krylov subspace method. We do not, however, pursue this aspect further, here. As another observation, let us note that if $\gamma_5 b = \pm b$, then computing $D_n b$ requires only one evaluation of the sign function instead of two ("chiral projection approach").

As with the other formulations, let us summarize the important spectral properties of the squared operator D_n.

Lemma 4. *With the notation of Lemma 1,* $\text{spec}(D_n) = \{\lambda_{j,\pm}^n\}$ *with*

$$\lambda_{j,\pm}^n = 1 + 2\phi_j \rho + \rho^2 \text{ (double eigenvalue)} \quad \text{if } \sigma_j \neq 0$$
$$\lambda_{j,+} = (\rho + \phi_j)^2, \; \lambda_{j,-} = (-\rho + \psi_j)^2 \quad \text{if } \sigma_j = 0.$$

The corresponding eigenvectors are the same as for D_h or, equivalently, the same as for D_u or, again equivalently,

$$\begin{pmatrix} x_j^1 \\ 0 \end{pmatrix}, \begin{pmatrix} 0 \\ x_j^2 \end{pmatrix}.$$

Notice also that $\text{spec}(D_n)$ satisfies

$$\text{spec}(D_n) \subseteq [(\rho-1)^2, (\rho+1)^2].$$

4 Optimal Krylov Subspace Methods

4.1 Propagator Computation

Let us start with the non-hermitian matrix D_u. Due to its shifted unitary form, there exists an optimal Krylov subspace method based on short recurrences to solve (2). This method was published in [8] and we would like to term it SUMR (shifted unitary minimal residual). SUMR is mathematically equivalent to full GMRES, so its residuals $r^m = b - D_u x^m$ at iteration m are minimal in the 2-norm in the corresponding affine Krylov subspace $x^0 + K_m(D_u, r^0)$ where $K_m(D_u, r^0) = \text{span}\{r^0, D_u r^0, \ldots, D_u^{m-1} r^0\}$. From the algorithmic point of view, SUMR is superior to full GMRES, since it relies on short recurrences and therefore requires constant storage and an equal amount of arithmetic work (one matrix vector multiplication and some vector operations) per iteration. The basic idea of SUMR is the observation that the upper Hessenberg matrix which describes the recursions of the Arnoldi process is shifted unitary so that its representation as a product of Givens rotations can be updated easily and with short recurrences. For the full algorithmic description we refer to [8].

Based on the spectral properties of D_u that we identified in Section 3, we can derive the following result on the convergence of SUMR for (2).

Lemma 5. *Let x^k be the k-th iterate of SUMR applied to (2) and let r_u^k be its residual. Then the following estimate holds:*

$$\|r_u^k\|_2 \leq 2 \cdot \left(\frac{1}{\rho}\right)^k \|r_u^0\|_2. \tag{12}$$

Proof. Since D_u is normal, its field of values $F(D_u) = \{\langle D_u x, x\rangle, \|x\|_2 = 1\}$ is the convex hull of its eigenvalues so that we have $F(D_u) \subseteq C(\rho, 1)$, the disk centered at ρ with radius 1. A standard result for full GMRES (which is mathematically equivalent to SUMR) now gives (12), see, e.g., [7].

Let us proceed with the hermitian operator D_h from (3), which is highly indefinite. The MINRES method is the Krylov subspace method of choice for such systems: It relies on short recurrences and it produces optimal iterates x^k in the sense that their residuals $r_h^k = \gamma_5 b - D_h x^k$ are minimal in the 2-norm over the affine subspace $x^0 + K_m(D_h, r_h^0)$.

Lemma 6. *Let x^k be the iterate of MINRES applied to (3) at iteration k and let r_h^k be its residual. Then the following estimate holds:*

$$\|r_h^k\|_2 \leq 2 \cdot \left(\frac{1}{\rho}\right)^{\lfloor k/2 \rfloor} \|r_h^0\|_2. \tag{13}$$

Here, $\lfloor k/2 \rfloor$ means $k/2$ rounded downwards to the nearest integer.

Proof. (13) is the standard MINRES estimate (see [7]) with respect to the information from (8).

As a last approach to solving the propagator equation, let us consider the standard normal equation

$$D_u D_u^\dagger z = b, \quad x = D_u^\dagger z. \tag{14}$$

Note that there exists an implementation of the CG method for (14) known as CGNE [6] which computes $x^k = D_u^\dagger z^k$ and its residual with respect to (2) on the fly, i.e., without additional work.

4.2 Dynamical Simulations

We now turn to the squared equation (9). Since D_n is hermitian and positive definite, the CG method is the method of choice for the solution of (9), its iterates y^m achieving minimal error in the energy norm (see Lemma 7 below) over the affine Krylov subspace $y^0 + K_m(D_n, r^0)$.

Lemma 7. *Let y^k be the iterate of CG applied to (9) at stage k. Then the following estimates hold ($y^* = D_n^{-1} b$)*

$$\|y^k - y^*\|_{D_n} \leq 2 \cdot \left(\frac{1}{\rho}\right)^k \|y^0 - y^*\|_{D_n}, \tag{15}$$

$$\|y^k - y^*\|_2 \leq 2 \cdot \frac{\rho+1}{\rho-1} \left(\frac{1}{\rho}\right)^k \|y^0 - y^*\|_2.$$

Here, $\|\cdot\|_{D_n}$ denotes the energy norm $\|y\|_{D_n} = \sqrt{y^\dagger D_n y}$.

Proof. The energy norm estimate (15) is the standard estimate for the CG method based on the bound $\mathrm{cond}(D_n) \leq ((\rho+1)/(\rho-1))^2$ for the condition number of D_n. The 2-norm estimate follows from the energy norm estimate using

$$\|y\|_2 \leq \sqrt{\|D_n^{-1}\|_2} \cdot \|y\|_{D_n} \leq \sqrt{\|D_n\|_2} \cdot \|y\|_2$$

with $\|D_n^{-1}\|_2 \leq 1/(\rho-1)^2$, $\|D_n\|_2 \leq (\rho+1)^2$.

5 Comparison of Methods

Based on Lemma 5 to 7 we now proceed by theoretically investigating the work for each of the three methods proposed so far. We consider two tasks: A propagator computation where we compute the solution x from (2) or (3), and a dynamical simulation where we need to solve (9).

5.1 Propagator Computation

The methods to be considered are SUMR for (2), MINRES for (3) and CGNE for (14). Note that due to (10), Lemma 7 can immediately also be applied to the CGNE iterates z^k which approximate the solution z of (14). In addition, expressing (15) in terms of $x^k = D_u^\dagger z^k$ instead of z^k turns energy norms into 2-norms, i.e. we have

$$\|x^k - x^*\|_2 \leq 2 \cdot \left(\frac{1}{\rho}\right)^k \|x^0 - x^*\|_2. \tag{16}$$

In order to produce a reasonably fair account of how many iterations we need, we fix a given accuracy ε for the final error and calculate the first iteration $k(\varepsilon)$ for which the results given in Lemmas 5 and 6 and in (16) guarantee

$$\| x^{k(\varepsilon)} - x^* \|_2 \leq \varepsilon \cdot \| r^0 \|_2, \quad x^* \text{ solution of (2)},$$

where $r^0 = b - D_u x^0$, x^0 being an identical starting vector for all three methods (most likely, $x^0 = 0$). Since $k(\varepsilon)$ will also depend on ρ, let us write $k(\varepsilon, \rho)$. The following may then be deduced from Lemma 5 and 6 and (16) in a straightforward manner.

Lemma 8. *(i) For SUMR we have*

$$\| x^k - x^* \|_2 \leq \frac{1}{\rho - 1} \cdot \| r_u^k \|_2 \leq \frac{2}{\rho - 1} \left(\frac{1}{\rho}\right)^k \cdot \| r^0 \|_2,$$

and therefore

$$k(\varepsilon, \rho) \leq \frac{-\ln(\varepsilon)}{\ln(\rho)} + \frac{-\ln(2/(\rho-1))}{\ln(\rho)}.$$

(ii) For MINRES we have using $\|r_h^0\| = \|r_u^0\|$, since $r_h^0 = \gamma_5 r_u^0$

$$\| x^k - x^* \|_2 \leq \frac{1}{\rho - 1} \cdot \| r_h^k \|_2 \leq \frac{2}{\rho - 1} \cdot \left(\frac{1}{\rho}\right)^{\lfloor \frac{k}{2} \rfloor} \cdot \| r_h^0 \|_2,$$

and therefore

$$k(\varepsilon, \rho) \leq 2 \cdot \left(\frac{-\ln(\varepsilon)}{\ln(\rho)} + \frac{-\ln(2/(\rho-1))}{\ln(\rho)}\right).$$

(iii) For CGNE we have

$$\| x^k - x^* \|_2 \leq 2 \cdot \left(\frac{1}{\rho}\right)^k \cdot \| x^0 - x^* \|_2 \leq 2 \cdot \left(\frac{1}{\rho}\right)^k \cdot \frac{1}{\rho - 1} \cdot \| r^0 \|_2,$$

and therefore

$$k(\varepsilon, \rho) \leq \frac{-\ln(\varepsilon)}{\ln(\rho)} + \frac{-\ln(2/(\rho - 1))}{\ln(\rho)}.$$

The arithmetic work in all these iterative methods is completely dominated by the cost for evaluating the matrix vector product $\text{sign}(Q)v$. MINRES and SUMR require one such evaluation per iteration, whereas CGNE requires two. Taking this into account, Lemma 8 suggests that MINRES and CGNE should require about the same work to achieve a given accuracy ε, whereas SUMR should need only half as much work, thus giving a preference for SUMR. Of course, such conclusions have to be taken very carefully: In a practical computation, the progress of the iteration will depend on the *distribution* of the eigenvalues, whereas the numbers $k(\varepsilon, \rho)$ of Lemma 1 were obtained by just using *bounds* for the eigenvalues. For large values of ρ the theoretical factor two between SUMR and MINRES/CGNE on the other hand, can be understood heuristically by the observation that SUMR can already reduce the residual significantly by placing one root of its corresponding polynomial in the center of the disc $C(\rho, 1)$ whereas for the hermitian indefinite formulation two roots are necessary in the two separate intervals. For smaller values of ρ the differences are expected to be smaller except for eigenvalues of D_u close to $\rho + 1$.

Figure 3 plots convergence diagrams for all three methods for our example configurations. The diagrams plot the relative norm of the residual $\|r^k\|/\|r^0\|$ as a function of the total number of matrix vector multiplications with the matrix Q. These matrix vector multiplications represent the work for evaluating $\text{sign}(Q)v$ in each iterative step, since we use the multishift CG method on a Zolotarev approximation with an accuracy of 10^{-8} with 10 poles (configuration A) and 20 poles (configuration B) respectively to approximate $\text{sign}(Q)v$. The true residual (dotted) converges to the accuracy of the inner iteration. Note that the computations for configuration B are much more demanding, since the evaluation of $\text{sign}(Q) \cdot v$ is more costly. We did not use any projection techniques to speed up this part of the computation.

Figure 4 plots convergence diagrams for all three methods for additional examples on a 8^4 lattice ($\beta = 5.6$, $\kappa = 0.2$, $\rho = 1.06$) and a 16^4 lattice ($\beta = 6.0$, $\kappa = 0.2$, $\rho = 1.06$) respectively.

MINRES and CGNE behave very similarly on configuration A. This is to be expected, since the hermitian indefinite matrix D_h is maximally indefinite, so that for an arbitrary right hand side the squaring of the matrix inherent in D_n should not significantly increase the number of required iterations.

SUMR always performs best. The savings compared to MINRES and CGNE depend on ρ, β and the lattice size and reached up to 50%.

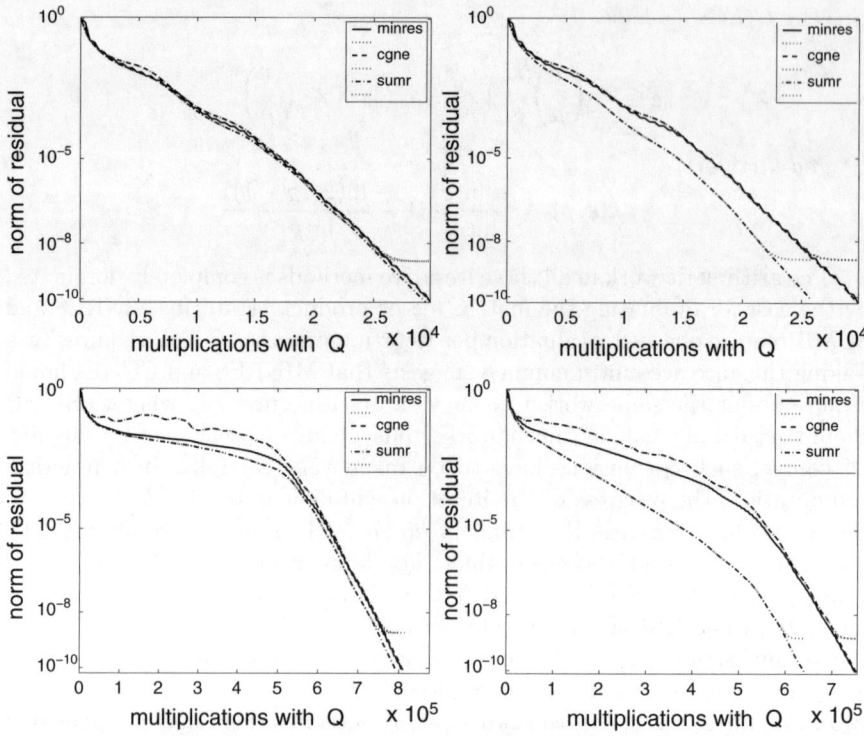

Fig. 3. Propagator computation: Convergence of MINRES for (3), CGNE for (14) and SUMR for (2). Left column is for $\rho = 1.01$, right column for $\rho = 1.1$. The upper plots are for configuration A, the lower for configuration B.

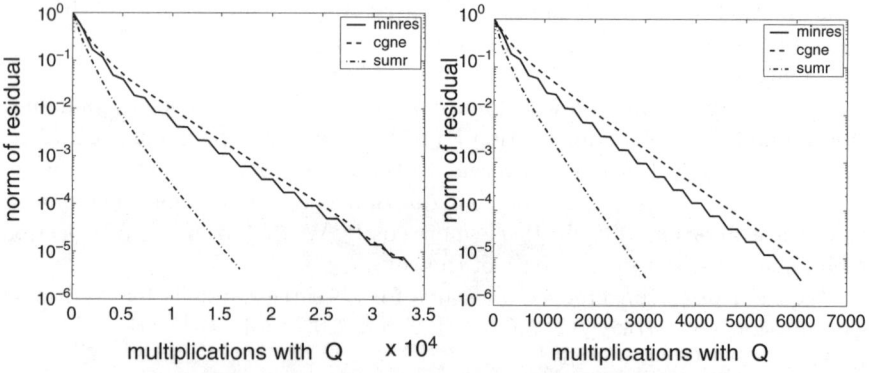

Fig. 4. Propagator computation: Convergence of MINRES for (3), CGNE for (14) and SUMR for (2). Left plot is for a 8^4 lattice, right plot for a 16^4 lattice.

As a side remark, let us mention that the dichotomy between hermitian indefinite and non-hermitian positive definite formulations is also a field of study in other areas. For example, [3] investigates the effect of multiplying some rows of a hermitian matrix with minus one in the context of solving augmented systems of the form

$$\begin{pmatrix} A & B \\ B^T & 0 \end{pmatrix} \begin{pmatrix} x \\ y \end{pmatrix} = \begin{pmatrix} b \\ 0 \end{pmatrix} \iff \begin{pmatrix} A & B \\ -B^T & 0 \end{pmatrix} \begin{pmatrix} x \\ y \end{pmatrix} = \begin{pmatrix} b \\ 0 \end{pmatrix}.$$

5.2 Dynamical Simulations

Let us now turn to a dynamical simulation where we compute the solution y^* from (9). The methods to be considered are CG for (9), a two-sweep SUMR-approach where we solve the two systems

$$D_u^\dagger x = b, \quad D_u y = x$$

using SUMR for both systems (note that $D_u^\dagger$ is of shifted unitary form, too), or a two-sweep MINRES-approach solving the two systems

$$D_h x = b, \quad D_h y = x.$$

It is now a bit more complicated to guarantee a comparable accuracy for each of the methods. Roughly speaking, we have to run both sweeps in the two-sweep methods to the given accuracy. Lemma 8 thus indicates that the two-sweep MINRES approach will not be competitive, whereas it does not determine which of two-sweep SUMR or CG is to be preferred. Our actual numerical experiments indicate that, in practice, CG is superior to two-sweep SUMR, see Figure 5.

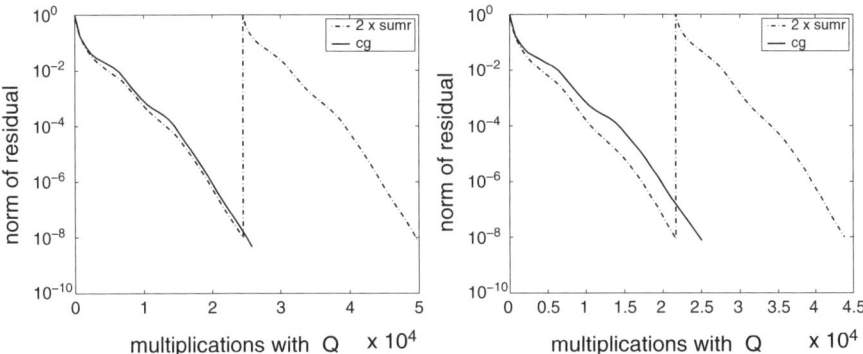

Fig. 5. Dynamical simulation: Convergence of two-sweep SUMR (dash-dotted) and CG (solid) for (9) Left plot is for $\rho = 1.01$, right plot for $\rho = 1.1$. Both plots are for configuration A.

6 Conclusion

We have for the first time applied SUMR as the outer iteration when solving a propagator equation (2) for overlap fermions. Our theoretical analysis and the numerical results indicate that this new method is superior to CGNE as well as to MINRES on the symmetrized equation. In practice, savings tend to increase with larger values of ρ. We achieved savings of about 50% for physically interesting configurations.

We have also shown that when solving the squared equation (9) directly, as is required in a dynamical simulation, a two sweep approach using SUMR is not competitive to directly using CG with the squared operator. This is in contrast to the case of the Wilson fermion matrix, where it was shown in [4] that the two-sweep approach using BiCGstab performs better than CG.[5]

In the context of the overall inner-outer iteration scheme, additional issues arise. In particular, one should answer the question of how accurately the result of the inner iteration (evaluating the product of sign(Q) with a vector) is really needed. This issue will be addressed in a forthcoming paper, [1].

Acknowledgements. G.A. is supported under Li701/4-1 (RESH Forschergruppe FOR 240/4-1). N.C. enjoys support from the EU Research and Training Network HPRN-CT-2000-00145 "Hadron Properties from Lattice QCD".

A Definitions

The Wilson-Dirac matrix reads $M = I - \kappa D_W$ where

$$(D_W)_{nm} = \sum_{\mu}(I - \gamma_\mu) \otimes U_\mu(n)\delta_{n,m-\mu} + (I + \gamma_\mu) \otimes U_\mu^\dagger(n-\mu)\delta_{n,m+\mu}. \quad (17)$$

The Euclidean γ-matrices in the chiral representation are given as:

$$\gamma_1 := \begin{bmatrix} 0 & 0 & i & 0 \\ 0 & 0 & 0 & i \\ -i & 0 & 0 & 0 \\ 0 & -i & 0 & 0 \end{bmatrix} \quad \gamma_2 := \begin{bmatrix} 0 & 0 & -1 & 0 \\ 0 & 0 & 0 & 1 \\ -1 & 0 & 0 & 0 \\ 0 & 1 & 0 & 0 \end{bmatrix} \quad \gamma_3 := \begin{bmatrix} 0 & 0 & 0 & i \\ 0 & 0 & -i & 0 \\ 0 & i & 0 & 0 \\ -i & 0 & 0 & 0 \end{bmatrix} \quad \gamma_4 := \begin{bmatrix} 0 & 0 & 0 & 1 \\ 0 & 0 & 1 & 0 \\ 0 & 1 & 0 & 0 \\ 1 & 0 & 0 & 0 \end{bmatrix}.$$

The hermitian form of the Wilson-Dirac matrix is given by

$$Q = \gamma_5 M, \quad (18)$$

with γ_5 defined as the product

$$\gamma_5 := \gamma_1\gamma_2\gamma_3\gamma_4 = \begin{bmatrix} 1 & 0 & 0 & 0 \\ 0 & 1 & 0 & 0 \\ 0 & 0 & -1 & 0 \\ 0 & 0 & 0 & -1 \end{bmatrix}. \quad (19)$$

---

[5] BiCGstab is not an alternative to SUMR in the case of the overlap operator, since, unlike BiCGstab, SUMR is an optimal method achieving minimal residuals in the 2-norm.

B Massive Overlap Operator

Following Neuberger [10], one can write the massive overlap operator as

$$D_N(\mu) = c\left((1+\mu)I + (1-\mu)M(M^\dagger M)^{-\frac{1}{2}}\right). \tag{20}$$

The normalisation c can be absorbed into the fermion renormalisation, and will not contribute to any physics. For convenience, we have set $c = 1/(1-\mu)$. Thus, the regularizing parameter ρ as defined in eq. (1) is related to μ by

$$\rho = (1+\mu)/(1-\mu). \tag{21}$$

The physical mass of the fermion is then given by

$$m_f = \frac{2\mu}{\kappa(1-\mu)}. \tag{22}$$

References

1. G. Arnold, N. Cundy, J. van den Eshof, A. Frommer, S. Krieg, Th. Lippert, and K. Schäfer. Numerical methods for the QCD overlap operator: III. nested iterations. *Comp. Phys. Comm.*, to appear.
2. R. G. Edwards, U. M. Heller, and R. Narayanan. A study of practical implementations of the overlap-Dirac operator in four dimensions. *Nucl. Phys.*, B540:457–471, 1999.
3. B. Fischer, A. Ramage, D. J. Silvester, and A. J. Wathen. Minimum residual methods for augmented systems. *BIT*, 38(3):527–543, 1998.
4. A. Frommer, V. Hannemann, B. Nöckel, Th. Lippert, and K. Schilling. Accelerating Wilson fermion matrix inversions by means of the stabilized biconjugate gradient algorithm. *Int. J. of Mod. Phys.* **C**, 5(6):1073–1088, 1994.
5. U. Glässner, S. Güsken, Th. Lippert, G. Ritzenhöfer, K. Schilling, and A. Frommer. How to compute Green's functions for entire mass trajectories within Krylov solvers. *Int. J. Mod. Phys.*, C7:635, 1996.
6. G. H. Golub and C. F. Van Loan. *Matrix Computations*. The Johns Hopkins University Press, Baltimore, London, 3rd edition, 1996.
7. A. Greenbaum. *Iterative Methods for Solving Linear Systems*, volume 17 of *Frontiers in Applied Mathematics*. Society for Industrial and Applied Mathematics (SIAM), Philadelphia, PA, 1997.
8. C. F. Jagels and L. Reichel. A fast minimal residual algorithm for shifted unitary matrices. *Numer. Linear Algebra Appl.*, 1(6):555–570, 1994.
9. R. Narayanan and H. Neuberger. An alternative to domain wall fermions. *Phys. Rev.*, D62:074504, 2000.
10. H. Neuberger. Vector like gauge theories with almost massless fermions on the lattice. *Phys. Rev.*, D57:5417–5433, 1998.
11. C. C. Paige and M. Wei. History and generality of the CS decomposition. *Linear Algebra Appl.*, 208/209:303–326, 1994.
12. J. van den Eshof, A. Frommer, Th. Lippert, K. Schilling, and H.A. van der Vorst. Numerical methods for the QCD overlap operator: I. sign-function and error bounds. *Comp. Phys. Comm.*, 146:203–224, 2002.

Fast Evaluation of Zolotarev Coefficients

A. D. Kennedy

School of Physics, University of Edinburgh, EH9 3JZ, Scotland, UK
adk@ph.ed.ac.uk

Summary. We review the theory of elliptic functions leading to Zolotarev's formula for the sign function over the range $\varepsilon \leq |x| \leq 1$. We show how Gauss' arithmetico-geometric mean allows us to evaluate elliptic functions cheaply, and thus to compute Zolotarev coefficients "on the fly" as a function of ε. This in turn allows us to calculate the matrix functions $\operatorname{sgn} H$, $\sqrt{H}$, and $1/\sqrt{H}$ both quickly and accurately for any Hermitian matrix H whose spectrum lies in the specified range.

1 Introduction

The purpose of this paper is to provide a detailed account of how to compute the coefficients of Zolotarev's optimal rational approximation to the sgn function. This is of considerable interest for lattice QCD because evaluation of the Neuberger overlap operator [1, 2, 3, 4] requires computation of the sgn function applied to a Hermitian matrix H. Numerical techniques for applying a rational approximation to a matrix are discussed in a companion paper [5], and in [6, 7].

In general, the computation of optimal (Chebyshev) rational approximations for a continuous function over a compact interval requires an iterative numerical algorithm [8, 9], but for the function $\operatorname{sgn} H$ (and the related functions $\sqrt{H}$ and $1/\sqrt{H}$ [5]) the coefficients of the optimal approximation are known in closed form in terms of Jacobi elliptic functions [10].

We give a fairly detailed summary of the theory of elliptic functions (§2) [11, 12] leading to the principal modular transformation of degree n (§2.7), which directly leads to Zolotarev's formula (§3). Our approach closely follows that presented in [11].

We also explain how to evaluate the elliptic functions necessary to compute the Zolotarev coefficients (§4.4), explaining the use of the appropriate modular transformations (§2.7) and of Gauss' arithmetico-geometric mean (§4.2), as well as providing explicit samples of code for the latter (§4.3).

2 Elliptic Functions

2.1 Introduction

There are two commonly encountered types of elliptic functions: Weierstrass (§2.4) and Jacobi (§2.6) functions. In principle these are completely equivalent: indeed each may be expressed in terms of the other (§2.6); but in practice Weierstrass functions are more elegant and natural for a theoretical discussion, and Jacobi functions are the more convenient for numerical use in most applications.

Elliptic functions are doubly periodic complex analytic functions (§2.2); the combination of their periodicity and analyticity leads to very strong constraints on their structure, and these constraints are most easily extracted by use of Liouville's theorem (§2.3). The constraints imply that for a fixed pair of periods ω and ω' an elliptic function is uniquely determined, up to an overall constant factor, by the locations of its poles and zeros. In particular, this means that any elliptic function may be expanded in *rational function* or *partial fraction* form in terms of the Weierstrass (§2.5) function with the same periods and its derivative. Furthermore, this in turn shows that all elliptic functions satisfy an *addition theorem* (§2.5) which allows us to write these expansions in terms of Weierstrass functions with unshifted argument z (§2.5).

There are many different choices of periods that lead to the same period lattice, and this representation theorem allows us to express them in terms of each other: such transformations are called *modular transformations of degree one*. We may also specify periods whose period lattice properly contains the original period lattice; and elliptic functions with these periods may be represented rationally in terms of the original ones. These (non-invertible) transformations are *modular transformations* of higher degree, and the set of all modular transformations form a semigroup that is generated by a few basic transformations (the Jacobi real and imaginary transformations, and the principal transformation of degree n, for instance). The form of these modular transformations may be found using the representation theorem by matching the location of the poles and zeros of the functions, and fixing the overall constant at some suitable position.

One of the periods of the Weierstrass functions may be eliminated by rescaling the argument, and if we accept this trivial transformation then all elliptic functions may be expressed in terms of the Jacobi functions (§2.6) with a single parameter k.

In order to evaluate the Jacobi functions for arbitrary argument and (real) parameter we may first use the Jacobi real transformation to write them in terms of Jacobi functions whose parameter lies in the unit interval; then we may use the addition theorem to write them in terms of functions of purely real and imaginary arguments, and finally use the Jacobi imaginary transformation to rewrite the latter in terms of functions with real arguments. This can all be done numerically or analytically, and is explained in detail for the case of interest in (§4).

We are left with the problem of evaluating Jacobi functions with real argument and parameter in the unit interval. This may be done very efficiently by use of Gauss' method of the arithmetico-geometric mean (§4.2). This makes use of the particular case of the principal modular transformation of degree 2, known as the *Gauss transformation* to show that the mapping $(a, b) \mapsto \left(\frac{1}{2}(a+b), \sqrt{ab}\right)$ iterates to a fixed point; for a suitable choice of the initial values the value of the fixed point gives us the value of the complete elliptic integral K, and with just a little more effort it can be induced to give us the values of all the Jacobi functions too (§4).

This procedure is sufficiently fast and accurate that the time taken to evaluate the coefficients of the Zolotarev approximation for any reasonable values of the range specified by ε and the degree n is negligible compared to the cost of applying the approximate operator $\operatorname{sgn} H$ to a vector.

2.2 Periodic Functions

A function $f: \mathbb{C} \to \mathbb{C}$ is *periodic* with *period* $\omega \in \mathbb{C}$ if $f(z) = f(z + \omega)$. Clearly, if $\omega_1, \omega_2, \ldots$ are periods of f then any linear combination of them with integer coefficients, $\sum_i n_i \omega_i$, is also a period; thus the periods form a $\mathbb{Z}$-module.

It is obvious that if f is a constant then this $\mathbb{Z}$-module is dense in $\mathbb{C}$, but the converse holds too, for if there is a sequence of periods $\omega_1, \omega_2, \ldots$ that converges to zero, then $f'(z) = \lim_{n \to \infty} [f(z + \omega_n) - f(z)]/\omega_n = 0$. It follows that every non-constant function must have a set of *primitive periods*, that is ones that are not sums of integer multiples of periods of smaller magnitude. Jacobi showed that if f is not constant it can have at most two primitive periods, and that these two periods cannot be collinear.

2.3 Liouville's Theorem

From here on we shall consider only doubly periodic meromorphic functions, which for historical reasons are called *elliptic functions*, whose non-collinear primitive periods we shall call ω and ω'. Consider the integral of such a function f around the parallelogram ∂P defined by its primitive periods,

$$\oint_{\partial P} dz\, f(z) = \int_0^\omega dz\, f(z) + \int_\omega^{\omega+\omega'} dz\, f(z) + \int_{\omega+\omega'}^{\omega'} dz\, f(z) + \int_{\omega'}^0 dz\, f(z).$$

Substituting $z' = z - \omega$ in the second integral and $z'' = z - \omega'$ in the third we have

$$\oint_{\partial P} dz\, f(z) = \int_0^\omega dz\, [f(z) - f(z+\omega')] + \int_0^{\omega'} dz\, [f(z+\omega) - f(z)] = 0,$$

upon observing that the integrands identically vanish due to the periodicity of f. On the other hand, since f is meromorphic we can evaluate it in terms of its residues, and hence we find that the sum of the residues at all the poles of f in P is zero. Since the sum of the residues at all the poles of an elliptic function are zero an elliptic function cannot have less than two poles, taking multiplicity into account.

Several useful corollaries follow immediately from this theorem. Consider the logarithmic derivative $g(z) = [\ln f(z)]' = f(z)'/f(z)$ where f is any elliptic function which is not identically zero. We see immediately that g is holomorphic everywhere except at the discrete set $\{\zeta_j\}$ where f has a pole or a zero. Near these singularities f has the Laurent expansion $f(z) = c_j(z - \zeta_j)^{r_j} + O\left((z - \zeta_j)^{r_j+1}\right)$ with $c_j \in \mathbb{C}$ and $r_j \in \mathbb{Z}$, so the residue of g at ζ_j is r_j. Applying the previous result to the function g instead of f we find that $\oint_{\partial P} dz\, g(z) = 2\pi i \sum_j r_j = 0$, or in other words that the number of poles of f must equal the number of zeros of f, counting multiplicity in both cases.

It follows immediately that there are no non-constant holomorphic elliptic functions; for if there was an analytic elliptic function f with no poles then $f(z) - a$ could have no zeros either.

If we consider the function $h(z) = zg(z)$ then we find

$$\oint_{\partial P} dz\, h(z) = \int_0^\omega dz\, h(z) + \int_\omega^{\omega+\omega'} dz\, h(z) + \int_{\omega+\omega'}^{\omega'} dz\, h(z) + \int_{\omega'}^0 dz\, h(z)$$

$$= \int_0^\omega dz\, [h(z) - h(z+\omega')] + \int_0^{\omega'} dz\, [h(z+\omega) - h(z)]$$

$$= \int_0^\omega dz\, [zg(z) - (z+\omega')g(z)] + \int_0^{\omega'} dz\, [(z+\omega)g(z) - zg(z)]$$

$$= -\omega' \int_0^\omega dz\, g(z) + \omega \int_0^{\omega'} dz\, g(z)$$

$$= -\omega'\{\ln[f(\omega)] - a] - \ln[f(0) - a]\} + \omega\{\ln[f(\omega') - a] - \ln[f(0) - a]\}$$

$$= 2\pi i(n'\omega' + n\omega),$$

where $n, n' \in \mathbb{N}$ are the number of times $f(z)$ winds around the origin as z is taken along the straight line from 0 to ω or ω'. On the other hand, Cauchy's theorem tells us that

$$\oint_{\partial P} dz\, h(z) = \oint_{\partial S} \frac{dz\, zf'(z)}{f(z)} = 2\pi i \sum_{k=1}^m (\alpha_k - \beta_k),$$

where α_k and β_k are the locations of the poles and zeros respectively of $f(z)$, again counting multiplicity. Consequently we have that $\sum_{k=1}^m (\alpha_k - \beta_k) = n\omega + n'\omega'$, that is, the sum of the locations of the poles minus the sum of the location of the zeros of any elliptic function is zero modulo its periods.

2.4 Weierstrass Elliptic Functions

The most elegant formalism for elliptic functions is due to Weierstrass. A simple way to construct a doubly periodic function out of some analytic function f is to construct the double sum $\sum_{m,m'\in\mathbb{Z}} f(z - m\omega - m'\omega')$. In order for this sum to converge uniformly it suffices that $|f(z)| < k/z^3$, so a simple choice is $Q(z) \equiv -2\sum_{m,m'\in\mathbb{Z}}(z - m\omega - m'\omega')^{-3}$. Clearly this function is doubly periodic, $Q(z+\omega) = Q(z+\omega') = Q(z)$, and odd, $Q(-z) = -Q(z)$.

The derivative of an elliptic function is clearly also an elliptic function, but in general the integral of an elliptic function is not an elliptic function. Indeed, if we define the Weierstrass $\wp$ function[1] such that $\wp' = Q$ we know that $\wp(z+\omega) = \wp(z)+c$ for any period ω. In this case we also know that $\wp$ must be an even function, $\wp(-z) = \wp(z)$, because Q is an odd function, and thus we have $\wp(\frac{1}{2}\omega) = \wp(-\frac{1}{2}\omega)+c$ by periodicity and $\wp(\frac{1}{2}\omega) = \wp(-\frac{1}{2}\omega)$ by symmetry, and hence $c = 0$. We have thus shown that

$$\wp(z) \equiv \frac{1}{z^2} + \int_0^z d\zeta \left\{ Q(\zeta) + \frac{2}{\zeta^3} \right\}$$

is an elliptic function. Its only singularities are a double pole at the origin and its periodic images.

---

[1] The name of the function is $\wp$, but I do not know what the name of the function is called; q.v., "Through the Looking-Glass, and what Alice found there," Chapter VIII, p. 306, footnote 8 [13].

If we expand $\wp$ in a Laurent series about the origin we obtain

$$\wp(z) = \frac{1}{z^2} + \sum_{j=1}^{\infty} \sum_{\substack{m,m' \in \mathbb{Z} \\ |m|+|m'| \neq 0}} \frac{(2j+1)z^{2j}}{(m\omega + m'\omega')^{2(j+1)}} \equiv \frac{1}{z^2} + \frac{g_2}{20}z^2 + \frac{g_3}{28}z^4 + \cdots,$$

where the coefficients are functions only of the periods

$$\frac{g_2}{60} = \sum_{\substack{m,m' \in \mathbb{Z} \\ |m|+|m'| \neq 0}} \frac{1}{(m\omega + m'\omega')^4}, \quad \text{and} \quad \frac{g_3}{140} = \sum_{\substack{m,m' \in \mathbb{Z} \\ |m|+|m'| \neq 0}} \frac{1}{(m\omega + m'\omega')^6}. \quad (1)$$

From this we find $\wp'(z) = -2z^{-3} + \frac{1}{10}g_2 z + \frac{1}{7}g_3 z^3 + \cdots$, and therefore $[\wp'(z)]^2 = 4z^{-6}\left\{1 - \frac{1}{10}g_2 z^4 - \frac{1}{7}g_3 z^6 + \cdots\right\}$ and $[\wp(z)]^3 = z^{-6}\left\{1 + \frac{3}{20}g_2 z^4 + \frac{3}{28}g_3 z^6 + \cdots\right\}$. Putting these together we find

$$[\wp'(z)]^2 - 4[\wp(z)]^3 + g_2 \wp(z) = -g_3 + Az^2 + Bz^4 + \cdots.$$

The left-hand side is an elliptic function with periods ω and ω' whose only poles are at the origin and its periodic images, the right-hand side has the value $-g_3$ at the origin, and thus by Liouville's theorem it must be a constant. We thus have $[\wp'(z)]^2 = 4[\wp(z)]^3 - g_2 \wp(z) - g_3$ as the differential equation satisfied by $\wp$. Indeed, this equation allows us to express all the derivatives of $\wp$ in terms of $\wp$ and $\wp'$; for example

$$\begin{aligned} \wp'' &= 6\wp^2 - \tfrac{1}{2}g_2, & \wp''' &= 12\wp\wp', \\ \wp^{(4)} &= 6(20\wp^3 - 3g_2\wp - 2g_3), & \wp^{(5)} &= 18(20\wp^2 - g_2)\wp'. \end{aligned} \quad (2)$$

We can formally solve the differential equation for $\wp$ to obtain the *elliptic integral* which is the functional inverse of $\wp$ (for fixed periods ω and ω'),

$$z = \int_0^z d\zeta = -\int_0^z \frac{d\zeta \, \wp'(\zeta)}{\sqrt{4\wp(\zeta)^3 - g_2 \wp(\zeta) - g_3}} = \int_{\wp(z)}^{\infty} \frac{dw}{\sqrt{4w^3 - g_2 w - g_3}}.$$

It is useful to factor the cubic polynomial which occurs in the differential equation, $\wp'^2(z) = 4(\wp - e_1)(\wp - e_2)(\wp - e_3)$, where the symmetric polynomials of the roots satisfy $e_1 + e_2 + e_3 = 0$, $e_1 e_2 + e_2 e_3 + e_3 e_1 = -\frac{1}{4}g_2$, $e_1 e_2 e_3 = \frac{1}{4}g_3$, and $e_1^2 + e_2^2 + e_3^2 = (e_1 + e_2 + e_3)^2 - 2(e_1 e_2 + e_2 e_3 + e_3 e_1) = \frac{1}{2}g_2$.

Since $\wp'$ is an odd function we have $-\wp'(\frac{1}{2}\omega) = \wp'(-\frac{1}{2}\omega) = \wp'(\frac{1}{2}\omega) = 0$, and likewise $\wp'(\frac{1}{2}\omega') = 0$ and $\wp'(\frac{1}{2}(\omega + \omega')) = 0$. The values of $\wp$ at the half-periods must be distinct, for if $\wp(\frac{1}{2}\omega) = \wp(\frac{1}{2}\omega')$ then the elliptic function $\wp(z) - \wp(\frac{1}{2}\omega')$ would have a double zero at $z = \frac{1}{2}\omega'$ and at $z = \frac{1}{2}\omega$, which would violate Liouville's theorem. Since the $\wp'$ vanishes at the half periods the differential equation implies that $\wp(\frac{1}{2}\omega) = e_1$, $\wp(\frac{1}{2}\omega') = e_2$, $\wp(\frac{1}{2}(\omega + \omega')) = e_3$, and that e_1, e_2, and e_3 are all distinct.

The solution of the corresponding differential equation with a generic quartic polynomial, $y'^2 = a(y - r_1)(y - r_2)(y - r_3)(y - r_4)$ with $r_i \neq r_j$, is easily found in terms of the Weierstrass function by a conformal transformation. First one root is mapped to infinity by the transformation $y = r_4 + 1/x$, giving $x'^2 = -a(x - \rho_1)(x - \rho_2)(x - \rho_3)/\rho_1 \rho_2 \rho_3$ with $\rho_j = 1/(r_j - r_4)$. Then the linear transformation $x = A\xi + B$ with $A = -4\rho_1 \rho_2 \rho_3/a$ and $B = (\rho_1 + \rho_2 + \rho_3)/3$ maps this to $\xi'^2 = 4(\xi - e_1)(\xi - e_2)(\xi - e_3)$

where $e_j = (\rho_j - B)/A$. The solution is thus $y = r_4 + 1/(A\wp + B)$, where $\wp$ has the periods implicitly specified by the roots e_j.

It is not obvious that there exist periods ω and ω' such that g_2 and g_3 are given by (1), nevertheless this is so (see [11] for a proof).

A simple example of this is given by the *Jacobi elliptic function* sn z, which is defined by $z \equiv \int_0^{\text{sn } z} dt \left[(1-t^2)(1-k^2t^2)\right]^{-\frac{1}{2}}$, and hence satisfies the differential equation $(\text{sn}'z)^2 = \left(1 + (\text{sn } z)^2\right)\left(1 - k^2(\text{sn } z)^2\right)$ together with the boundary condition $\text{sn } 0 = 0$. We may move one of the roots to infinity by substituting $\text{sn } z = 1 + 1/x(z)$ and multiplying through by $x(z)^4$, giving $x'^2 = -2(1 - k^2)\left[x + \frac{1}{2}\right]\left[x - k/(1-k)\right]\left[x + k/(1+k)\right]$. The linear change of variable $x(z) = -\left[12\xi(z) + 1 - 5k^2\right]/\left[6(1-k^2)\right]$ then puts this into Weierstrass's canonical form $\xi'^2 = 4(\xi - e_1)(\xi - e_2)(\xi - e_3) = 4\xi^3 - g_2\xi - g_3$ with the roots

$$e_1 = \frac{k^2 + 1}{6}, \quad e_{\frac{5\pm1}{2}} = -\frac{k^2 \pm 6k + 1}{12}; \tag{3}$$

and correspondingly $g_2 = \frac{1}{12}(k^4 + 14k^2 + 1)$, and $g_3 = \frac{1}{6^3}(k^2+1)(k^2+6k+1)(k^2-6k+1)$. Clearly the Weierstrass function $\wp(z)$ with periods corresponding to the roots e_j is a solution to this equation. A more general solution may be written as $\xi(z) = \wp(f(z))$ for some analytic function f; for this to be a solution it must satisfy the differential equation, which requires that $f'(z)^2 = 1$, so $\xi(z) = \wp(\pm z + \Delta)$ with Δ a suitable constant chosen to satisfy the boundary conditions. It turns out that the boundary values required for sn are satisfied by the choice $\xi(z) = \wp(z - K(k))$, where $K(k) \equiv \int_0^1 dt \left[(1-t^2)(1-k^2t^2)\right]^{-\frac{1}{2}}$ is the *complete elliptic integral*. We shall later derive the expression for sn in terms of the Weierstrass functions $\wp$ and $\wp'$ with the same argument and periods by a simpler method.

The Weierstrass ζ-Function

It is useful to consider integrals of $\wp$, even though these are not elliptic functions. If we define $\zeta' = -\wp$, whose solution is

$$\zeta(z) \equiv \frac{1}{z} - \int_0^z du \left\{\wp(u) - \frac{1}{u^2}\right\}$$

where the path of integration avoids all the singularities of $\wp$ (i.e., the periodic images of the origin) except for the origin itself. The only singularities of ζ are a simple pole with unit residue at the origin and its periodic images. Furthermore, ζ is an odd function. However, ζ is not periodic: $\zeta(z+\omega) = \zeta(z) + \eta$ where η is a constant. Setting $z = -\frac{1}{2}\omega$ and using the fact that ζ is odd we obtain $\zeta(-\frac{1}{2}\omega) = \zeta(\frac{1}{2}\omega) + \eta = -\zeta(\frac{1}{2}\omega)$, or $\zeta(\frac{1}{2}\omega) = \frac{1}{2}\eta$. If we integrate ζ around a period parallelogram P containing the origin we find a useful identity relating ω, ω', η and η': $2\pi i = \oint_{\partial P} dz\, \zeta(z) = \int_c^{c+\omega} dz\, [\zeta(z) - \zeta(z+\omega')] + \int_c^{c+\omega'} dz\, [\zeta(z+\omega) - \zeta(z)] = \int_c^{c+\omega'} dz\, \eta - \int_c^{c+\omega} dz\, \eta' = \eta\omega' - \eta'\omega$.

The Weierstrass σ-Function

That was so much fun that we will do it again. Let $(\ln \sigma)' = \sigma'/\sigma = \zeta$, so

$$\sigma(z) \equiv z \exp\left[\int_0^z du \left\{\zeta(u) - \frac{1}{u}\right\}\right],$$

where again the integration path avoids all the singularities of ζ except the origin. σ is a holomorphic function having only simple zeros lying at the origin and its periodic images, and it is odd. To find the values of σ on the period lattice we integrate $\sigma'(z+\omega)/\sigma(z+\omega) = \zeta(z+\omega) = \zeta(z) + \eta = \sigma'(z)/\sigma(z) + \eta$ to obtain $\ln\sigma(z+\omega) = \ln\sigma(z) + \eta z + c$, or $\sigma(z+\omega) = c'e^{\eta z}\sigma(z)$. As usual we can find the constant c' by evaluating this expression at $z = -\frac{1}{2}\omega$, $\sigma(\frac{1}{2}\omega) = -c'e^{-\frac{1}{2}\eta\omega}\sigma(\frac{1}{2}\omega)$, giving $c' = -e^{\frac{1}{2}\eta\omega}$ and $\sigma(z+\omega) = -e^{\eta(z+\frac{1}{2}\omega)}\sigma(z)$.

2.5 Expansion of Elliptic Functions

Every rational function $R(z)$ can be expressed in two canonical forms, either in a fully factored representation which makes all the poles and zeros explicit,

$$R(z) = c\frac{(z-b_1)(z-b_2)\cdots(z-b_n)}{(z-a_1)(z-a_2)\cdots(z-a_m)},$$

or in a partial fraction expansion which makes the leading "divergent" part of its Laurent expansion about its poles manifest,

$$R(z) = E(z) + \sum_{i,k} \frac{A_k^{(i)}}{(z-a_k)^i}.$$

In these expressions b_i are the zeros of R, a_i its poles, c and $A_k^{(i)}$ are constants, and E is a polynomial. It is perhaps most natural to think of E, the *entire part* of R, as the leading terms of its Laurent expansion about infinity.

An arbitrary elliptic function f with periods ω and ω' may be expanded in two analogous ways in terms of Weierstrass elliptic functions with the same periods.

Multiplicative Form

To obtain the first representation recall that $\sum_{j=1}^n (a_j - b_j) = 0 \pmod{\omega, \omega'}$, so we can choose a set of poles and zeros (not necessarily in the fundamental parallelogram) whose sum is zero. For instance, we could just add the appropriate integer multiples of ω and ω' to a_1. We now construct the function

$$g(z) = \frac{\sigma(z-b_1)\sigma(z-b_2)\cdots\sigma(z-b_n)}{\sigma(z-a_1)\sigma(z-a_2)\cdots\sigma(z-a_n)},$$

which has the same zeros and poles as f. Furthermore, it is also an elliptic function, since $g(z+\omega) = \exp\left[\eta\sum_{j=1}^n (a_j - b_j)\right] g(u) = g(u)$. It follows that the ratio $f(u)/g(u)$ is an elliptic function with no poles, as for each pole in the numerator there is a corresponding pole in the denominator, and for each zero in the denominator there is a corresponding zero in the numerator. Therefore, by Liouville's theorem, the ratio must be a constant $f(u)/g(u) = C$, so we have

$$f(z) = C\frac{\sigma(z-b_1)\sigma(z-b_2)\cdots\sigma(z-b_n)}{\sigma(z-a_1)\sigma(z-a_2)\cdots\sigma(z-a_n)}.$$

Additive Form

For the second "partial fraction" representation let $a_1, \ldots, a_n$ be the poles of f lying in some fundamental parallelogram. In this case, unlike the previous one, we ignore multiplicity and count each pole just once in this list. Further, let the leading terms of the Laurent expansion about $z = a_k$ be $\sum_{r=1}^{m_k} (-1)^r (r-1)! A_k^{(r-1)} (z-a_k)^{-r}$; the function $g_k(z) \equiv -\sum_{r=1}^{m_k} A_k^{(r-1)} \zeta^{(r-1)}(z-a_k) = -A_k^{(0)} \zeta(z-a_k) + \sum_{r=2}^{m_k} A_k^{(r-1)} \wp^{(r-2)}(z-a_k)$ then has exactly the same leading terms in its Laurent expansion.

Summing this expression over all the poles, we obtain $g(z) = \sum_{k=1}^n g_k(z) = -\sum_{k=1}^n \sum_{r=1}^{m_k} A_k^{(r-1)} \zeta^{(r-1)}(z - a_k)$. The sum of the terms with $r > 1$, being sums of the elliptic function $\wp(z - a_k)$ and its derivatives, is an elliptic function. The sum of terms with $r = 1$, $\varphi(z) = -\sum_{k=1}^n A_k^{(0)} \zeta(z - a_k)$, behaves under translation by a period as $\varphi(z + \omega) = \varphi(z) - \eta \sum_{k=1}^n A_k^{(0)} = \varphi(z)$, where we have used the corollary of Liouville's theorem that the sum of the residues at all the poles of the elliptic function f in a fundamental parallelogram is zero. It follows that the sum of $r = 1$ terms is an elliptic function also, so the difference $f(z) - g(z)$ is an elliptic function with no singularities, and thus by Liouville's theorem is a constant C. We have thus obtain the expansion of an arbitrary elliptic function $f(z) = C + g(z) = C - \sum_{k=1}^n \sum_{r=1}^{m_k} A_k^{(r-1)} \zeta^{(r-1)}(z - a_k)$, where the ζ functions have the same periods as f does.

Addition Theorems

Consider the elliptic function $f(u) = \wp'(u) / (\wp(u) - \wp(v))$; according to Liouville's theorem the denominator must have exactly two simple zeros, at which $\wp(u) = \wp(v)$, within any fundamental parallelogram. $\wp$ is an even function, $\wp(-v) = \wp(v)$, so these zeros occur at $u = \pm v$. At $u = 0$ the function f has a simple pole, and the leading terms of the Laurent series for f about these three poles is $(u-v)^{-1} + (u+v)^{-1} - 2/u$. The "partial fraction" expansion of f is thus $f(u) = C + \zeta(u-v) + \zeta(u+v) - 2\zeta(u)$, and since both f and ζ are odd functions we observe that $C = 0$.

Adding this result, $\wp'(u)/(\wp(u) - \wp(v)) = \zeta(u-v) + \zeta(u+v) - 2\zeta(u)$, to the corresponding equation with u and v interchanged, $-\wp'(v)/(\wp(u) - \wp(v)) = -\zeta(u-v) + \zeta(u+v) - 2\zeta(v)$, gives $(\wp'(u) - \wp'(v))/(\wp(u) - \wp(v)) = 2\zeta(u+v) - 2\zeta(u) - 2\zeta(v)$. Rearranging this gives the *addition theorem* for zeta functions $\zeta(u + v) = \zeta(u) + \zeta(v) + \frac{1}{2}(\wp'(u) - \wp'(v))/(\wp(u) - \wp(v))$.

The corresponding addition theorem for $\wp$ is easily obtained by differentiating this relation

$$-\wp(u+v) = -\wp(u) + \frac{1}{2} \frac{\wp''(u)[\wp(u) - \wp(v)] - \wp'(u)[\wp'(u) - \wp'(v)]}{[\wp(u) - \wp(v)]^2}$$

and adding to it the same formula with u and v interchanged to obtain

$$-2\wp(u+v) = -\wp(u) - \wp(v) + \frac{1}{2} \frac{[\wp''(u) - \wp''(v)][\wp(u) - \wp(v)] - [\wp'(u) - \wp'(v)]^2}{[\wp(u) - \wp(v)]^2}.$$

Recalling that a consequence of the differential equation satisfied by $\wp$ is the identity (2), $2\wp'' = 12\wp^2 - g_2$, we have $\wp''(u) - \wp''(v) = 6[\wp(u)^2 - \wp(v)^2]$, and thus $\wp(u+v) = -\wp(u) - \wp(v) + \frac{1}{4}\left(\frac{\wp'(u) - \wp'(v)}{\wp(u) - \wp(v)}\right)^2$.

Differentiating this addition theorem for $\wp$ gives the addition theorem for $\wp'$. Since higher derivatives of $\wp$ can be expressed in terms of $\wp$ and $\wp'$ there is no need to repeat this construction again.

Representation of elliptic functions in terms of $\wp$ and $\wp'$

Consider the "partial fraction" expansion of an arbitrary elliptic function f in terms of zeta functions and their derivatives, $f(z) = C + \sum_{k=1}^{n} \sum_{r=1}^{m_k} A_k^{(r-1)} \zeta^{(r-1)}(z-a_k)$. This expresses $f(z)$ as a linear combination of zeta functions $\zeta(z-a_k)$, which are not elliptic functions, and their derivatives $\zeta^{(r)}(z-a_k) = -\wp^{(r-1)}(z-a_k)$ which are. We may now use the addition theorems to write this in terms of the zeta function $\zeta(z)$ and its derivatives $\zeta^{(r)}(z) = -\wp^{(r-1)}(z)$ of the unshifted argument z.

For the $r = 1$ terms the zeta function addition theorem gives us $\sum_{k=1}^{n} A_k^{(0)} \zeta(z - a_k) = \sum_{k=1}^{n} A_k^{(0)} \zeta(z) + R_1(\wp(z), \wp'(z))$, were we use the notation $R_i(x, y)$ to denote a rational function of x and y; i.e., an element of the field $\mathbb{C}(x, y)$. The coefficients in R_1 depend transcendentally on a_k, of course. This expression simplifies to just the rational function $R_1(\wp, \wp')$ on recalling that, as we have previously shown, $\sum_{k=1}^{n} A_k^{(0)} = 0$.

Using the addition theorems for $\wp$ and $\wp'$ all the terms for $r > 1$ may be expressed in the form $\sum_{k=1}^{n} A_k^{(r-1)} \zeta^{(r-1)}(z - a_k) = -\sum_{k=1}^{n} A_k^{(r-1)} \wp^{(r-2)}(z - a_k) = R_r(\wp(z), \wp'(z))$. We have thus shown that $f = R(\wp, \wp')$. In fact, since the differential equation for $\wp$ expresses $\wp'^2$ as a polynomial in $\wp$ we can simplify this to the form $f = R_e(\wp) + R_o(\wp)\wp'$. A simple corollary is that if f is an even function then $f = R_e(\wp)$ and if it is odd then $f = R_o(\wp)\wp'$.

A corollary of this result is that *any two elliptic functions with the same periods are algebraic functions of each other.* If f and g are two such functions then $f = R_1(\wp) + R_2(\wp)\wp'$, $g = R_3(\wp) + R_4(\wp)\wp'$, and $\wp'^2 = 4\wp^3 - g_2\wp - g_3$; these equations immediately give three polynomial relations between the values f, g, $\wp$, and $\wp'$, and we may eliminate the last two to obtain a polynomial equation $F(f, g) = 0$. To be concrete, suppose $R_i(z) = P_i(z)/Q_i(z)$ with $P_i, Q_i \in \mathbb{C}[z]$, then we have

$$\bar{P}_1(f, \wp, \wp') \equiv Q_1(\wp)Q_2(\wp)f - P_1(\wp)Q_2(\wp) - P_2(\wp)Q_1(\wp)\wp' = 0,$$
$$\bar{P}_2(g, \wp, \wp') \equiv Q_3(\wp)Q_4(\wp)g - P_3(\wp)Q_4(\wp) - P_4(\wp)Q_3(\wp)\wp' = 0,$$
$$\bar{P}_3(\wp, \wp') \equiv \wp'^2 - 4\wp^3 + g_2\wp + g_3 = 0;$$

we may then construct the resultants[2]

$$\bar{P}_4(f, \wp) \equiv \operatorname*{Res}_{\wp'}\left(\bar{P}_1(f, \wp, \wp'), \bar{P}_3(\wp, \wp')\right) = 0,$$
$$\bar{P}_5(g, \wp) \equiv \operatorname*{Res}_{\wp'}\left(\bar{P}_2(g, \wp, \wp'), \bar{P}_3(\wp, \wp')\right) = 0,$$
$$F(f, g) \equiv \operatorname*{Res}_{\wp}\left(\bar{P}_4(f, \wp), \bar{P}_5(g, \wp)\right) = 0.$$

A corollary of this corollary is obtained by letting $g = f'$, which tells us that every elliptic function satisfies a first order differential equation of the form $F(f, f') = 0$ with $F \in \mathbb{C}[f, f']$.

---

[2] In practice Gröbner basis methods might be preferred.

A second metacorollary is obtained by considering $g(u) = f(u+v)$, for which we deduce that there is a polynomial $\mathbb{C}\langle v\rangle[f(u)][f(u+v)] \ni F = 0$, where $\mathbb{C}\langle v\rangle$ is the space of complex-valued transcendental functions of v. On the other hand, interchanging u and v we observe that $F \in \mathbb{C}\langle u\rangle[f(v)][f(u+v)]$ too. The coefficients of F are therefore both polynomials in $f(u)$ with coefficients which are functions of v, and polynomials in $f(v)$ with coefficients which are functions of u. It therefore follows that the coefficients must be polynomials in $f(u)$ and $f(v)$ with constant coefficients, $F \in \mathbb{C}[f(u), f(v), f(u+v)]$. In other words, every elliptic equation has an algebraic addition theorem.

2.6 Jacobi Elliptic Functions

We shall now consider the Jacobi elliptic function sn implicitly defined by $z \equiv \int_0^{\operatorname{sn} z} dt\, [(1-t^2)(1-k^2 t^2)]^{-\frac{1}{2}}$. This cannot be anything new — it must be expressible rationally in terms of the Weierstrass functions $\wp$ and $\wp'$ with the same periods.

The integrand of the integral defining sn has a two-sheeted Riemann surface with four branch points. The values of z for which $\operatorname{sn}(z,k) = s$ for any particular value $s \in \mathbb{C}$ are specified by the integral; we immediately see that there two such values, corresponding to which sheet of the integrand we end up on, plus arbitrary integer multiples of the two periods ω and ω'. These periods correspond to the non-contractible loops that encircle any pair of the branch points. There are only two independent homotopically non-trivial loops because the contour which encircles all four branch points is contractible through the point at infinity (this is a regular point of the integrand, as may be seen by changing variable to $1/z$).

We may choose the first period to correspond to a contour C which contains the branch points at $z = \pm 1$. We find

$$\omega = \oint_C \frac{dt}{\sqrt{(1-t^2)(1-k^2t^2)}} = \left[\int_0^1 - \int_1^0 - \int_0^{-1} + \int_{-1}^0\right] \frac{dt}{\sqrt{(1-t^2)(1-k^2t^2)}} = 4K(k),$$

taking into account the fact that the integrand changes sign as we go round each branch point onto the other sheet of the square root. Likewise, we may choose the second period to correspond to a contour C' enclosing the branch points at $z=1$ and $z=1/k$; this gives

$$\omega' = \oint_{C'} \frac{dt}{\sqrt{(1-t^2)(1-k^2t^2)}} = \left[\int_1^{1/k} - \int_{1/k}^1\right] \frac{dt}{\sqrt{(1-t^2)(1-k^2t^2)}}.$$

If we change variable to $s = \sqrt{(1-k^2t^2)/(1-k^2)}$ we find that

$$\int_1^{1/k} \frac{dt}{\sqrt{(1-t^2)(1-k^2t^2)}} = i\int_0^1 \frac{ds}{\sqrt{(1-s^2)(1-k'^2 s^2)}} = iK(k'),$$

where we define $k' \equiv \sqrt{1-k^2}$. We thus have shown that the second period $\omega' = 2iK(k')$ is also expressible as a complete elliptic integral.

The locations of the poles of sn are also easily found. Consider the integral $\int_{1/k}^\infty dt\,[(1-t^2)(1-k^2t^2)]^{-\frac{1}{2}}$, by the change of variable $s = 1/kt$ we see that it is equal to $\int_0^1 ds\,(ks^2)^{-1}\left[(1-1/k^2 s^2)(1-1/s^2)\right]^{-\frac{1}{2}} = K(k)$. We therefore have that

$$\int_0^\infty \frac{dt}{\sqrt{(1-t^2)(1-k^2t^2)}} = \left[\int_0^1 + \int_1^{1/k} + \int_{1/k}^\infty\right] \frac{dt}{\sqrt{(1-t^2)(1-k^2t^2)}} = 2K(k) + iK(k'),$$

and thus sn has a pole at $2K(k) + iK(k')$.

We mentioned that there are always two locations within the fundamental parallelogram at which $\operatorname{sn}(z, k) = s$. One of these locations corresponds to a contour C_1 which goes from $t = 0$ on the principal sheet (the positive value of the square root in the integrand) to $t = s$ on the same sheet, while the other goes from $t = 0$ on the principal sheet to $t = s$ on the second sheet. This latter contour is homotopic to one which goes from $t = 0$ on the principal sheet to $t = 0$ on the second sheet and then follows C_1 but on the second sheet. If the value of the first integral is z, then the value of the second is $2K(k) - z$, thus establishing the identity $\operatorname{sn}(z, k) = \operatorname{sn}(2K(k) - z, k)$.

Since the integrand is an even function of t the integral is an odd function of sn, from which we immediately see that $\operatorname{sn}(z, k) = -\operatorname{sn}(-z, k)$.

We summarise these results by giving some of the values of sn within the fundamental parallelogram defined by $\omega = 4K$ and $\omega' = 2iK'$:

z	0	K	$2K$	$3K$	iK'	$K + iK'$	$2K + iK'$	$3K + iK'$
$\operatorname{sn}(z, k)$	0	1	0	-1	∞	$1/k$	$-\infty$	$-1/k$

where we have used the notation $K \equiv K(k)$ and $K' \equiv K(k')$.

Representation of sn in terms of $\wp$ and $\wp'$

From this knowledge of the periods, zeros, and poles of sn we can express it in terms of Weierstrass elliptic functions. From (3) we know that the periods $\omega = 4K$, $\omega' = 2iK'$, and $\omega + \omega' = 4K + 2iK'$ correspond to the roots e_1, e_2, and e_3; that is $\wp\left(\frac{1}{2}\omega\right) = e_1$, $\wp\left(\frac{1}{2}\omega'\right) = e_2$, and $\wp\left(\frac{1}{2}(\omega + \omega')\right) = e_3$. Since $\operatorname{sn}(z, k)$ is an odd function of z it must be expressible as $R(\wp(z))\wp'(z)$ where R is a rational function; since it has simple poles in the fundamental parallelogram only at $z = \frac{1}{2}\omega' = iK'$ and $z = \frac{1}{2}(\omega + \omega') = 2K + iK'$ it must be of the form $\operatorname{sn}(z, k) = \bar{R}(\wp(z))\wp'(z)/[(\wp(z) - e_2)(\wp(z) - e_3)]$ with $\bar{R}$ a rational function.[3] The Weierstrass function has a double pole at the origin, its derivative has a triple pole, and the Jacobi elliptic function sn has a simple zero, so we can deduce that $\bar{R}(\wp(z))$ must be regular and non-zero at the origin, and hence $\bar{R}$ is just a constant. As $\operatorname{sn}(z, k) = z + O(z^3)$ near the origin this constant is easily determined by considering the residues of the poles in the Weierstrass functions, and we obtain the interesting identity $\operatorname{sn}(z, k) = -\frac{1}{2}\wp'(z)/[(\wp(z) - e_2)(\wp(z) - e_3)]$.

Representation of sn² in terms of $\wp$

We can use the same technique to express $\operatorname{sn}(z, k)^2$ in terms of Weierstrass elliptic functions. The differential equation satisfied by $s(z) \equiv \operatorname{sn}(z, k)^2$ is $s'^2 = 4s(1-s)(1-k^2s)$, which is reduced to Weierstrass canonical form $\zeta'^2 = 4\zeta^3 - \bar{g}_2\zeta - \bar{g}_3$ with $\bar{g}_2 = \frac{4}{3}(k^4 + k^2 + 1)$ and $\bar{g}_3 = \frac{4}{27}(2k^2 - 1)(k^2 - 2)(k^2 + 1)$ by the linear substitution $s = \left(\zeta + \frac{1}{3}(1 + k^2)\right)/k^2$. The roots of this cubic form are $\bar{e}_1 = -\frac{1}{3}(1 + k^2)$, $\bar{e}_{\frac{5 \pm 1}{2}} =$

---
[3] Remember that the logarithmic derivative of a function $d\ln f(z)/dz = f'(z)/f(z)$ always has a simple pole at each pole and zero of f.

$\frac{1}{3}(2 \pm k^2)$. The general solution of this equation is $\zeta(z) = \wp(\pm z + c)$ where c is a constant, and since $\text{sn}(z,k)^2$ has a double pole at $z = iK(k')$ we have $\text{sn}(z,k)^2 = \left[\wp\left(z - iK(k')\right) + \frac{1}{3}(1+k^2)\right]/k^2$. This can be simplified using the addition formula for Weierstrass elliptic functions to give[4] $\text{sn}(z,k)^2 = 1/\left[\wp(z) - \bar{e}_1\right]$. Of course, we could have seen this immediately by noting that $\text{sn}(z,k)^2$ is an even elliptic function with periods $2K(k)$ and $2iK(k')$ corresponding to the roots $\bar{e}_i$, and therefore must be a rational function of $\wp(z)$. Since it has a double pole at $z = iK(k')$ and a double zero at $z = 0$ in whose neighbourhood $\text{sn}(z,k)^2 = z^2 + O(z^4)$ the preceding expression is uniquely determined.

A useful corollary of this result is that we can express the Weierstrass function $\wp(z)$ with periods $2K(k)$ and $2iK'(k)$ rationally in terms of $\text{sn}(z,k)^2$, namely $\wp(z) = \text{sn}(z,k)^{-2} + \bar{e}_1$, and thus any even elliptic function with these periods may be written as a rational function of $\text{sn}(z,k)^2$.

Addition formula for Jacobi elliptic functions

We may derive the explicit addition formula for Jacobi elliptic functions using a method introduced by Euler. Consider the functions $s_1 \equiv \text{sn}(u,k)$, $s_2 \equiv \text{sn}(v,k)$ where we shall hold $u + v = c$ constant. The differential equations for s_1 and s_2 are $s_1'^2 = (1-s_1^2)(1-k^2s_1^2)$, $s_2'^2 = (1-s_2^2)(1-k^2s_2^2)$, where we have used a prime to indicate differentiation with respect to u and noted that $v' = -1$. Multiplying the equations by s_2^2 and s_1^2 respectively and subtracting them gives $W(s_1,s_2) \cdot (s_1 s_2)' = (s_1 s_2' - s_2 s_1')(s_1 s_2' + s_2 s_1') = (s_1^2 - s_2^2)(1 - k^2 s_1^2 s_2^2)$, where we have introduced the Wronskian $W(s_1, s_2) \equiv \det \begin{pmatrix} s_1 & s_2 \\ s_1' & s_2' \end{pmatrix}$. If we differentiate the differential equations for s_1 and s_2 we obtain $s_1'' = -(1+k^2)s_1 + 2k^2 s_1^3$, $s_2'' = -(1+k^2)s_2 + 2k^3 s_2^3$; subtracting these equations gives $W' = (s_1 s_2' - s_2 s_1')' = s_1 s_2'' - s_2 s_1'' = -2k^2 s_1 s_2(s_1^2 - s_2^2) = (s_1^2 - s_2^2)(1 - k^2 s_1^2 s_2^2)'/(s_1 s_2)'$. We may combine the expressions we have derived for W and W' to obtain $(\ln W)' = W'/W = (1-k^2 s_1^2 s_2^2)'/(1-k^2 s_1^2 s_2^2) = \left(\ln(1 - k^2 s_1^2 s_2^2)\right)'$. Upon integration this yields an explicit expression for the Wronskian, $W = C(1 - k^2 s_1^2 s_2^2)$ where C is a constant, by which we mean that it does not depend upon u. The constant does depend on the value of $c = u + v$, and it may be found by evaluating formula at $v = 0$.

To do so it is convenient to introduce two other Jacobi elliptic functions $\text{cn}(u,k) \equiv \sqrt{1 - \text{sn}(u,k)^2}$ where $\text{cn}(0,k) = 1$; and $\text{dn}(u,k) \equiv \sqrt{1 - k^2 \text{sn}(u,k)^2}$, where $\text{dn}(0,k) = 1$. In terms of these functions we may write $\text{sn}' u = \text{cn}\,u\,\text{dn}\,u$, furthermore they satisfy the identities $(\text{sn}\,u)^2 + (\text{cn}\,u)^2 = 1$ and $(k^2 \text{sn}\,u)^2 + (\text{dn}\,u)^2 = 1$, and differentiating these identities yields $\text{cn}'\,u = -\text{sn}\,u\,\text{dn}\,u$, $\text{dn}'\,u = -k^2 \text{sn}\,u\,\text{cn}\,u$.

We may now write $C = W(\text{sn}\,u, \text{sn}\,v)/\left[1 - (k\,\text{sn}\,u\,\text{sn}\,v)^2\right] = [\text{sn}\,u\,\text{cn}\,v\,\text{dn}\,v + \text{sn}\,v\,\text{cn}\,u\,\text{dn}\,u]/\left[1 - (k\,\text{sn}\,u\,\text{sn}\,v)^2\right]$, remembering that $v' = -1$. Setting $v = 0$ gives $C = \text{sn}\,u = \text{sn}\,c$, and thus we have the desired addition formula $\text{sn}(u+v) = (\text{sn}\,u\,\text{cn}\,v\,\text{dn}\,v + \text{sn}\,v\,\text{cn}\,u\,\text{dn}\,u) / \left(1 - (k\,\text{sn}\,u\,\text{sn}\,v)^2\right)$.

---

[4]The Weierstrass functions in this expression implicitly correspond to the roots $\bar{e}_i$, whereas as those in the previous expression for $\text{sn}(z,k)$ corresponded to the e_i.

2.7 Transformations of Elliptic Functions

So far we have studied the dependence of elliptic functions on their argument for fixed values of the periods. Although the Weierstrass function appear to depend on two arbitrary complex periods ω and ω' they really only depend on the ratio $\tau = \omega'/\omega$. If we rewrite the identity $\wp(z) = \wp(z+\omega) = \wp(z+\omega')$ in terms of the new variable $\zeta \equiv z/\omega$ we have $\wp(\omega\zeta) = \wp(\omega(\zeta+1)) = \wp(\omega(\zeta+\tau))$. Viewed as a function of ζ we have an elliptic function with periods 1 and τ, and as we have shown this is expressible rationally in terms of the Weierstrass function and its derivative with these periods.

Another observation is that there are many choices of periods ω and ω' which generate the same period lattice. Indeed, if we choose periods $\tilde{\omega} = \alpha\omega + \beta\omega'$, $\tilde{\omega}' = \gamma\omega + \delta\omega'$ with $\det\begin{pmatrix}\alpha & \beta \\ \gamma & \delta\end{pmatrix} = 1$ then this will be the case. This induces a relation between elliptic functions with these periods called a *first degree transformation*.

Jacobi imaginary transformation

Jacobi's imaginary transformation, or the second principal[5] first degree transformation, corresponds to the interchange of periods $\omega' = -\tilde{\omega}$ and $\omega = \tilde{\omega}'$. We start with the function $\operatorname{sn}(z,k)^2$ which has periods $\tilde{\omega} = 2K$ and $\tilde{\omega}' = 2iK'$, and consider the function $\operatorname{sn}(z/M,\lambda)^2$ with periods $\omega = 2ML$ and $\omega' = 2iML'$ (with $L = K(\lambda)$ and $L' = K(\lambda')$ as usual). For suitable M and λ we have $ML = iK'$ and $iML' = -K$, corresponding to the desired interchange of periods.

Since $\operatorname{sn}(z/M,\lambda)^2$ is an even function whose period lattice is the same as that of $\operatorname{sn}(z,k)^2$ it must be expressible as a rational function of $\operatorname{sn}(z,k)^2$, and this rational function may be found by matching the location of poles and zeros. $\operatorname{sn}(z/M,\lambda)^2$ has a double zero at $z/M = 0$ and a double pole at $z/M = iL'$. This latter condition may be written as $z = iML' = -K$, or $z = K$ upon using the periodicity conditions to map the pole into the fundamental parallelogram. Thus

$$\operatorname{sn}\left(\frac{z}{M},\lambda\right)^2 = \frac{A\operatorname{sn}(z,k)^2}{\operatorname{sn}(z,k)^2 - \operatorname{sn}(K,k)^2} = \frac{A\operatorname{sn}(z,k)^2}{\operatorname{sn}(z,k)^2 - 1}.$$

The constant A may be found by evaluating both sides of this equation at $z = iK'$: on the left $\operatorname{sn}(iK'/M,\lambda)^2 = \operatorname{sn}(L,\lambda)^2 = 1$, whereas on the right we have A because $\operatorname{sn}(z,k) \to \infty$ as $z \to iK'$. We thus have $A = 1$.

The value of λ is found by evaluating both sides at $z = -K + iK'$: on the left $\operatorname{sn}((-K+iK')/M,\lambda)^2 = \operatorname{sn}(iL' + L,\lambda)^2 = 1/\lambda^2$, and on the right we have $A/(1 - k^2)$ since $\operatorname{sn}(-K + iK',k)^2 = 1/k^2$. We thus have $\lambda = \sqrt{1-k^2} = k'$.

From these values for A and λ we may easily find M, as $iK' = ML = MK(\lambda) = MK(k') = MK'$ gives $M = i$. We may therefore write the Jacobi imaginary transformation as $\operatorname{sn}(-iz,k')^2 = \operatorname{sn}(z,k)^2/\left(\operatorname{sn}(z,k)^2 - 1\right)$, or equivalently $\operatorname{sn}(iz,k') = i\operatorname{sn}(z,k)/\operatorname{cn}(z,k)$, where we have made use of the fact that sn^2 is an even function, and chosen the sign of the square root according to the definition of cn and the fact that $\operatorname{sn}(z,k) = z + O(z^3)$.

---

[5] The first principal first degree transformation may be derived similarly. See [11] for details.

Principal transformation of degree n

We can also choose periods $\tilde{\omega}$ and $\tilde{\omega}'$ whose period lattice has the original one as a sublattice, for instance we may choose $\tilde{\omega} = \omega$ and $\tilde{\omega}' = \omega'/n$ where $n \in \mathbb{N}$. Elliptic functions with these periods must be rationally expressible in terms of the Weierstrass elliptic functions with the original periods, although the inverse may not be true. This relationship is called a *transformation of degree n*.

Let us construct such an elliptic function with periods $4K$ and $2iK'/n$, where $K \equiv K(k)$ and $K' \equiv K(k')$ with $k^2 + k'^2 = 1$. We may do this by taking $\operatorname{sn}(z,k)$ and scaling z by some factor $1/M$ and choosing a new parameter λ. We are thus led to consider the function $\operatorname{sn}(z/M, \lambda)$, whose periods with respect to z/M are $4L \equiv 4K(\lambda)$ and $2iL' \equiv 2iK(\lambda')$, with $\lambda^2 + \lambda'^2 = 1$. Viewed as a function of z it has periods $4LM$ and $2iL'M$, and M and λ are thus fixed by the conditions that $LM = K$ and $L'M = K'/n$. The ratio $f(z) \equiv \operatorname{sn}(z/M, \lambda)/\operatorname{sn}(z,k)$ must therefore be an even function of z with periods $2K$ and $2K'$;

$$f(\pm z + 2mK + 2im'K') = \frac{\operatorname{sn}\left(\frac{\pm z + 2mK + 2im'K'}{M}, \lambda\right)}{\operatorname{sn}(\pm z + 2mK + 2im'K', k)}$$

$$= \frac{\operatorname{sn}\left(\pm\frac{z}{M} + 2mL + 2im'nL', \lambda\right)}{\operatorname{sn}(\pm z + 2mK + 2im'K', k)} = \frac{\pm(-)^m \operatorname{sn}\left(\frac{z}{M}, \lambda\right)}{\pm(-)^m \operatorname{sn}(z,k)} = f(z)$$

for $m, m' \in \mathbb{Z}$; and hence $f(z)$ must be a rational function of $\operatorname{sn}(z,k)^2$.

Within its fundamental parallelogram the numerator of $f(z)$ has simple zeros at $z = m2iL'M = m2iK'/n$ for $m = 0, 1, \ldots, n-1$ and simple poles for $m = \frac{1}{2}, \frac{3}{2}, \ldots, n-\frac{1}{2}$; whereas its denominator has a simple zero at $z = 0$ and a simple pole at $z = iK'$. Hence, if n is even $f(z)$ has simple zeros for $m = 1, 2, \ldots, \frac{1}{2}n - 1, \frac{1}{2}n + 1, \ldots, n-1$, a double zero for $m = \frac{1}{2}n$, and simple poles for $m = \frac{1}{2}, \frac{3}{2}, \ldots, n - \frac{1}{2}$; whereas if n is odd then $m = 1, 2, \ldots, n-1$ give simple zeros and $m = \frac{1}{2}, \frac{3}{2}, \ldots, \frac{1}{2}n - 1, \frac{1}{2}n + 1, \ldots, n - \frac{1}{2}$ simple poles. Therefore there are $2\lfloor \frac{1}{2}n \rfloor$ zeros and poles, and it is easy to see that they always come in pairs such that the zeros occur for $\operatorname{sn}(z,k) = \pm\operatorname{sn}(2iK'm/n, k)$ and the poles for $\operatorname{sn}(z,k) = \pm\operatorname{sn}\left(2iK'(m-\frac{1}{2})/n, k\right)$ with $m = 1, \ldots, \lfloor \frac{1}{2}n \rfloor$. We thus see that the rational representation is

$$f(z) \equiv \frac{\operatorname{sn}\left(\frac{z}{M}, \lambda\right)}{\operatorname{sn}(z,k)} = \frac{1}{M} \prod_{m=1}^{\lfloor \frac{1}{2}n \rfloor} \frac{1 - \frac{\operatorname{sn}(z,k)^2}{\operatorname{sn}(2iK'm/n,k)^2}}{1 - \frac{\operatorname{sn}(z,k)^2}{\operatorname{sn}\left(2iK'(m-\frac{1}{2})/n,k\right)^2}}, \tag{4}$$

where the overall factor is determined by considering the behaviour near $z = 0$.

The value of the quantity M may be determined by evaluating this expression at the half period K where $\operatorname{sn}(K, k) = 1$ and $\operatorname{sn}(K/M, \lambda) = \operatorname{sn}(L, \lambda) = 1$, so

$$M = \prod_{m=1}^{\lfloor \frac{1}{2}n \rfloor} \frac{1 - \frac{1}{\operatorname{sn}(2iK'm/n,k)^2}}{1 - \frac{1}{\operatorname{sn}\left(2iK'(m-\frac{1}{2})/n,k\right)^2}}.$$

The value of the parameter λ is found by evaluating the identity at $z = K + iK'/n$, where $\operatorname{sn}\left(\frac{K+iK'/n}{M}, \lambda\right) = \operatorname{sn}\left(\frac{K}{M} + i\frac{K'}{nM}, \lambda\right) = \operatorname{sn}(L + iL', \lambda) = \frac{1}{\lambda}$.

It will prove useful to write the identity in parametric form

$$\operatorname{sn}\left(\frac{z}{M}, \lambda\right) = \frac{\xi}{M} \prod_{m=1}^{\lfloor \frac{1}{2}n \rfloor} \frac{1 - c_m \xi^2}{1 - c'_m \xi^2}, \tag{5}$$

with $\xi \equiv \operatorname{sn}(z, k)$, $c_m \equiv \operatorname{sn}(2iK'm/n, k)^{-2}$ and $c'_m \equiv \operatorname{sn}(2iK'(m - \frac{1}{2})/n, k)^{-2}$. This emphasises the fact that $\operatorname{sn}(z/M, \lambda)$ is a rational function of $\operatorname{sn}(z, k)$ of degree $(2\lfloor \frac{1}{2}n \rfloor + 1, 2\lfloor \frac{1}{2}n \rfloor)$.

3 Zolotarev's Problem

Zolotarev's fourth problem is to find the best uniform rational approximation to $\operatorname{sgn} x \equiv \vartheta(x) - \vartheta(-x)$ over the interval $[-1, -\varepsilon] \cup [\varepsilon, 1]$. This is easily done using the identity (5) derived in the preceding section.

We note that the function $\xi = \operatorname{sn}(z, k)$ with $k < 1$ is real and increases monotonically in $[0, 1]$ for $z \in [0, K]$, where as before we define $K \equiv K(k)$ to be a complete elliptic integral. Similarly we observe that $\operatorname{sn}(z, k)$ is real and increases monotonically in $[1, 1/k]$ for $z = K + iy$ with $y \in [0, K']$ and $K' \equiv K(k')$, $k^2 + k'^2 = 1$. On the other hand, $\operatorname{sn}(z/M, \lambda)$ has the same real period $2K$ as $\operatorname{sn}(z, k)$ and has an imaginary period $2iK'/n$ which divides that of $\operatorname{sn}(z, k)$ exactly n times. This means that $\operatorname{sn}(z/M, \lambda)$ also increases monotonically in $[0, 1]$ for $z \in [0, K]$, and then oscillates in $[1, 1/\lambda]$ for $z = K + iy$ with $y \in [0, K']$.

In order to produce an approximation of the required type we just need to rescale both the argument ξ so it ranges between -1 and 1 rather than $-1/k$ and $1/k$, and the function so that it oscillates symmetrically about 1 for $\xi \in [1, 1/k]$ rather than between 1 and $1/\lambda$. We thus obtain

$$R(x) = \frac{2}{1 + \frac{1}{\lambda}} \frac{x}{kM} \prod_{m=1}^{\lfloor \frac{1}{2}n \rfloor} \frac{k^2 - c_m x^2}{k^2 - c'_m x^2} \tag{6}$$

with $k = \varepsilon$. On the domain $[-1, -\varepsilon] \cup [\varepsilon, 1]$ the error $e(x) \equiv R(x) - \operatorname{sgn}(x)$ satisfies $|e(x)| \leq \Delta \equiv \frac{1-\lambda}{1+\lambda}$, or in other words $\|R - \operatorname{sgn}\|_\infty = \Delta$. Furthermore, the error alternates $4\lfloor \frac{1}{2}n \rfloor + 2$ times between the extreme values of $\pm \Delta$, so by Chebyshev's theorem on optimal rational approximation R is the best rational approximation of degree $(2\lfloor \frac{1}{2}n \rfloor + 1, 2\lfloor \frac{1}{2}n \rfloor)$. In fact we observe that R is *deficient*, as its denominator is of degree one lower than this; this must be so as we are approximating an odd function. Indeed, we may note that $R'(x) \equiv (1 - \Delta^2)/R(x)$ is also an optimal rational approximation.

4 Numerical Evaluation of Elliptic Functions

We wish to consider Gauss' arithmetico-geometric mean as it provides a good means of evaluating Jacobi elliptic functions numerically.

4.1 Gauss Transformation

Gauss considered the transformation that divides the second period of an elliptic function by two, $\omega'_1 = \omega_1$ and $\omega'_2 = \frac{1}{2}\omega_2$. This is a special case of the principal transformation of nth degree on the second period considered before (4) with $n = 2$, hence

$$\operatorname{sn}\left(\frac{z}{M}, \lambda\right) = \frac{\operatorname{sn}(z,k)}{M}\left[\frac{1 - \frac{\operatorname{sn}(z,k)^2}{\operatorname{sn}(iK',k)^2}}{1 - \frac{\operatorname{sn}(z,k)^2}{\operatorname{sn}(\frac{1}{2}iK',k)^2}}\right],$$

with the parameter λ corresponding to periods $L = K/M$ and $L' = K'/2M$. Using Jacobi's imaginary transformation (the second principal first degree transformation, with $\omega_1' = -\omega_2$ and $\omega_2' = \omega_1$),

$$\operatorname{sn}(iz, k) = i\frac{\operatorname{sn}(z, k')}{\operatorname{cn}(z, k')}, \qquad (7)$$

we get

$$\operatorname{sn}\left(\frac{z}{M}, \lambda\right) = \frac{\operatorname{sn}(z,k)}{M}\left[\frac{1 + \frac{\operatorname{sn}(z,k)^2 \operatorname{cn}(K',k')^2}{\operatorname{sn}(K',k')^2}}{1 + \frac{\operatorname{sn}(z,k)^2 \operatorname{cn}(\frac{1}{2}K',k')^2}{\operatorname{sn}(\frac{1}{2}K',k')^2}}\right].$$

Since $\operatorname{sn}(K', k') = 1$, $\operatorname{cn}(K', k') = 0$,[6] $\operatorname{sn}(\frac{1}{2}K', k') = 1/\sqrt{1+k}$, and $\operatorname{cn}(\frac{1}{2}K', k') = \sqrt{k/(1+k)}$, we obtain $\operatorname{sn}(z/M, \lambda) = \operatorname{sn}(z, k)\left[1 + k\operatorname{sn}(z, k)^2\right]^{-1}/M$.

To determine M we set $z = K$: $1 = \operatorname{sn}(K/M, \lambda) = \operatorname{sn}(K, k)\left[1 + k\operatorname{sn}(K, k)^2\right]^{-1}/M = [1 + k]^{-1}/M$, hence $M = 1/(1+k)$. To determine λ we set $u = K + iK'/2$:

$$\operatorname{sn}\left(\frac{K}{M} + \frac{iK'}{2M}, \lambda\right) = \frac{\operatorname{sn}(K + \frac{1}{2}iK', k)}{M[1 + k\operatorname{sn}(K + \frac{1}{2}iK', k)^2]}.$$

Now, from the addition formula[7] $\operatorname{sn}(u + v) = (\operatorname{sn} u \operatorname{cn} v \operatorname{dn} v + \operatorname{sn} v \operatorname{cn} u \operatorname{dn} u)/[1 - (k\operatorname{sn} u \operatorname{sn} v)^2]$, we deduce that

$$\operatorname{sn}\left(K + \frac{iK'}{2}\right) = \frac{\operatorname{sn} K \operatorname{cn}\frac{iK'}{2}\operatorname{dn}\frac{iK'}{2} + \operatorname{sn}\frac{iK'}{2}\operatorname{cn} K \operatorname{dn} K}{1 - \left(k\operatorname{sn} K \operatorname{sn}\frac{iK'}{2}\right)^2} = \frac{\operatorname{cn}\frac{iK'}{2}\operatorname{dn}\frac{iK'}{2}}{1 - \left(k\operatorname{sn}\frac{iK'}{2}\right)^2}.$$

Furthermore $\operatorname{sn}(\frac{1}{2}iK', k) = i\operatorname{sn}(\frac{1}{2}K', k')/\operatorname{cn}(\frac{1}{2}K', k') = i/\sqrt{k}$, and correspondingly $\operatorname{cn}(\frac{1}{2}iK', k) = \sqrt{(1+k)/k}$, and $\operatorname{dn}(\frac{1}{2}iK', k) = \sqrt{1+k}$, giving $\operatorname{sn}(K + \frac{1}{2}iK', k) = 1/\sqrt{k}$. We thus find $1/\lambda = \operatorname{sn}(L + iL', \lambda) = 1/2M\sqrt{k}$ or $\lambda = 2M\sqrt{k} = 2\sqrt{k}/(1+k)$. Combining these results we obtain an explicit expression for Gauss' transformation

$$\operatorname{sn}\left((1+k)z, \frac{2\sqrt{k}}{1+k}\right) = \frac{(1+k)\operatorname{sn}(z, k)}{1 + k\operatorname{sn}(z, k)^2}.$$

---

[6]Let $x \equiv \operatorname{sn}(\frac{1}{2}K, k)$, then by the addition formula for Jacobi elliptic functions $\operatorname{sn}(K, k) = 1 = 2x\sqrt{1-x^2}\sqrt{1-k^2x^2}/(1-k^2x^4)$. Hence $(1 - k^2x^4)^2 = 4x^2(1-x^2)(1-k^2x^2)$, so $k^4x^8 - 4k^2x^6 + 2(2+k^2)x^4 - 4x^2 + 1 = 0$ or, with $z \equiv 1/x^2 - 1$, $[z^2 - (1-k^2)]^2 = 0$. Thus $z = \pm\sqrt{1-k^2} = \pm k'$, or $x = 1/\sqrt{1 \pm k'}$. Since $0 < x < 1$ we must choose the positive sign, so $\operatorname{sn}\left(\frac{1}{2}K(k), k\right) = 1/\sqrt{1+k'}$.

[7]We shall suppress the parameter k when it is the same for all the functions occurring in an expression.

4.2 Arithmetico-Geometric Mean

Let $a_n, b_n \in \mathbb{R}$ with $a_n > b_n > 0$, and define their arithmetic and geometric means to be $a_{n+1} \equiv \frac{1}{2}(a_n + b_n)$, $b_{n+1} \equiv \sqrt{a_n b_n}$. Since these are means we easily see that $a_n > a_{n+1} > b_n$ and $a_n > b_{n+1} > b_n$; furthermore $a_{n+1}^2 - b_{n+1}^2 = \frac{1}{4}(a_n^2 + 2a_n b_n + b_n^2) - a_n b_n = \frac{1}{4}(a_n^2 - 2a_n b_n + b_n^2) = \frac{1}{4}(a_n - b_n)^2 > 0$, so $a_n > a_{n+1} > b_{n+1} > b_n$. Thus the sequence converges to the *arithmetico-geometric mean* $a_\infty = b_\infty$.

If we choose a_n and b_n such that $k = (a_n - b_n)/(a_n + b_n)$, e.g., $a_n = 1 + k$ and $b_n = 1 - k$, then

$$1 + k = \frac{(a_n + b_n) + (a_n - b_n)}{a_n + b_n} = \frac{a_n}{a_{n+1}},$$

$$k^2 = \left(\frac{a_n - b_n}{a_n + b_n}\right)^2 = \frac{(a_n + b_n)^2 - 4a_n b_n}{(a_n + b_n)^2} = 1 - \frac{b_{n+1}^2}{a_{n+1}^2},$$

$$\frac{4k}{(1+k)^2} = \frac{(1+k)^2 - (1-k)^2}{(1+k)^2} = 1 - \left(\frac{1-k}{1+k}\right)^2$$

$$= 1 - \left(\frac{(a_n + b_n) - (a_n - b_n)}{(a_n + b_n) + (a_n - b_n)}\right)^2 = 1 - \frac{b_n^2}{a_n^2}.$$

If we define $s_n \equiv \operatorname{sn}\left((1+k)z, \frac{2\sqrt{k}}{1+k}\right)$ and $s_{n+1} \equiv \operatorname{sn}(z, k)$ then Gauss' transformation tells us that

$$s_n = \frac{(1+k)s_{n+1}}{1 + k s_{n+1}^2} = \frac{a_n s_{n+1}}{a_{n+1}\left[1 + \left(\frac{a_n - b_n}{a_n + b_n}\right)s_{n+1}^2\right]} = \frac{2a_n s_{n+1}}{(a_n + b_n) + (a_n - b_n)s_{n+1}^2}.$$

On the other hand

$$z = \int_0^{s_{n+1}} \frac{dt}{\sqrt{(1-t^2)(1-k^2 t^2)}} = \frac{1}{1+k} \int_0^{s_n} \frac{dt}{\sqrt{(1-t^2)\left[1 - \frac{4k}{(1+k)^2} t^2\right]}},$$

and these two integrals may be rewritten as

$$z = \int_0^{s_{n+1}} \frac{dt}{\sqrt{(1-t^2)\left[1 - \left(1 - \frac{b_{n+1}^2}{a_{n+1}^2}\right)t^2\right]}} = \frac{a_{n+1}}{a_n} \int_0^{s_n} \frac{dt}{\sqrt{(1-t^2)\left[1 - \left(1 - \frac{b_n^2}{a_n^2}\right)t^2\right]}}.$$

Therefore the quantity

$$\frac{z}{a_{n+1}} = \int_0^{s_{n+1}} \frac{dt}{\sqrt{(1-t^2)[a_{n+1}^2(1-t^2) + b_{n+1}^2 t^2]}} = \int_0^{s_n} \frac{dt}{\sqrt{(1-t^2)[a_n^2(1-t^2) + b_n^2 t^2]}}$$

is invariant under the transformation $(a_n, b_n, s_n) \mapsto (a_{n+1}, b_{n+1}, s_{n+1})$, and thus

$$\frac{z}{a_{n+1}} = \int_0^{s_\infty} \frac{dt}{\sqrt{(1-t^2)[a_\infty^2(1-t^2) + b_\infty^2 t^2]}} = \frac{1}{a_\infty} \int_0^{s_\infty} \frac{dt}{\sqrt{1-t^2}} = \frac{\sin^{-1} s_\infty}{a_\infty}.$$

This implies that $s_\infty = \sin\left(\frac{a_\infty z}{a_{n+1}}\right) = \sin(a_\infty z)$ with our previous choice of $a_n = 1 + k, b_n = 1 - k \Rightarrow a_{n+1} = 1, b_{n+1} = \sqrt{1 - k^2}$. We may thus compute $s_{n+1} = \operatorname{sn}(z, k) = f(z, 1, \sqrt{1-k^2})$ for $0 < k < 1$ where

$$f(z, a, b) \equiv \begin{cases} \sin(az) & \text{if } a = b \\ \frac{2a\xi}{(a+b)+(a-b)\xi^2} \text{ with } \xi \equiv f\left(z, \frac{a+b}{2}, \sqrt{ab}\right) & \text{if } a \neq b \end{cases}$$

Furthermore, if we take $z = K(k)$ then $s_{n+1} = 1$ and $s_n = 2a_n/[(a_n + b_n) + (a_n - b_n)] = 1$; thus $s_\infty = \sin(a_\infty K) = 1$, so $a_\infty K = \pi/2$ or $K(k) = \pi/2a_\infty$.

4.3 Computer Implementation

An implementation of this method is shown in Figures 1 and 2.

The function `arithgeom` recursively evaluates the function f defined above. One subtlety is the stopping criterion, which has to be chosen carefully to guarantee that the recursion will terminate (which does not happen if the simpler criterion `b==a` is used instead) and which ensures that the solution is as accurate as possible whatever floating point precision `FLOAT` is specified. Another subtlety is how the value of the arithmetico-geometric mean `*agm` is returned from the innermost level of the recursion. Ideally, we would like this value to be bound to an automatic variable in the calling procedure `sncndnK` rather than passed as an argument, thus avoiding copying its address for every level of recursion (as is done in here) or copying its value for every level if it were explicitly returned as a value. Unfortunately this is impossible, since the C programming language does not allow us to have nested procedures. The reason we have written it in the present form is so that the code is thread-safe: if we made `agm` a static global variable then two threads simultaneously invoking `sncndnK` might interfere with each other's value. The virtue of this approach is only slightly tarnished by the fact that the global variable `pb` used in the convergence test is likewise not thread-safe. The envelope routine `sncndnK` is almost trivial, except that care is needed to get the sign of $\operatorname{cn}(z, k)$ correct.

4.4 Evaluation of Zolotarev Coefficients

The arithmetico-geometric mean lets us evaluate Jacobi elliptic functions for real arguments z and real parameters $0 < k < 1$. For complex arguments we can use the addition formula to evaluate $\operatorname{sn}(x + iy, k)$ in terms of $\operatorname{sn}(x, k)$ and $\operatorname{sn}(iy, k)$, and the latter case with an imaginary argument may be rewritten in terms of real arguments using Jacobi's imaginary transformation. We can either use these transformations to evaluate elliptic functions of complex argument numerically, or to transform algebraically the quantities we wish to evaluate into explicitly real form. Here we shall follow the latter approach, as it is more efficient to apply the transformations once analytically.

Zolotarev's formula (6) is

$$R(x) = \frac{2}{1 + \frac{1}{\lambda}} \frac{x}{kM} \prod_{m=1}^{\lfloor \frac{1}{2}n \rfloor} \frac{k^2 - c_m x^2}{k^2 - c'_m x^2}$$

```
#include <math.h>
#define ONE ((FLOAT) 1)
#define TWO ((FLOAT) 2)
#define HALF (ONE/TWO)

static void sncndnK(FLOAT z, FLOAT k, FLOAT* sn, FLOAT* cn,
                    FLOAT* dn, FLOAT* K) {
  FLOAT agm;
  int sgn;
  *sn = arithgeom(z, ONE, sqrt(ONE - k*k), &agm);
  *K = M_PI / (TWO * agm);
  sgn = ((int) (fabs(z) / *K)) % 4;  /* sgn = 0, 1, 2, 3 */
  sgn ^= sgn >> 1;    /* (sgn & 1) = 0, 1, 1, 0 */
  sgn = 1 - ((sgn & 1) << 1); /* sgn = 1, -1, -1, 1 */
  *cn = ((FLOAT) sgn) * sqrt(ONE - *sn * *sn);
  *dn = sqrt(ONE - k*k* *sn * *sn);
}
```

Fig. 1. The procedure sncndnK computes $\text{sn}(z, k)$, $\text{cn}(z, k)$, $\text{dn}(z, k)$, and $K(k)$. It is essentially a wrapper for the routine arithgeom shown in Figure 2. The sign of $\text{cn}(z, k)$ is defined to be -1 if $K(k) < z < 3K(k)$ and $+1$ otherwise, and this sign is computed by some quite unnecessarily obfuscated bit manipulations.

```
static FLOAT arithgeom(FLOAT z, FLOAT a, FLOAT b, FLOAT* agm) {
  static FLOAT pb = -ONE;
  FLOAT xi;

  if (b <= pb) { pb = -ONE; *agm = a; return sin(z * a); }
  pb = b;
  xi = arithgeom(z, HALF*(a+b), sqrt(a*b), agm);
  return 2*a*xi / ((a+b) + (a-b)*xi*xi);
}
```

Fig. 2. Recursive implementation of Gauss' arithmetico-geometric mean, which is the kernel of the method used to compute the Jacobi elliptic functions with parameter k where $0 < k < 1$. The function returns a value related to $\text{sn}(z, k')$, and also sets the value of *agm to the arithmetico-geometric mean. This value is simply related to complete elliptic function $K(k')$ and also determines the sign of $\text{cn}(z, k')$. The algorithm is deemed to have converged when b ceases to increase: this works whatever floating point precision FLOAT is specified.

with $c_m \equiv \text{sn}(2iK'm/n, k)^{-2}$ and $c'_m \equiv \text{sn}(2iK'(m-\frac{1}{2})/n, k)^{-2}$. We may evaluate the coefficients c_m and c'_m by using Jacobi's imaginary transformation (7),

$$c_m = -\left[\frac{\text{cn}(2K'm/n, k')}{\text{sn}(2K'm/n, k')}\right]^2, \quad c'_m = -\left[\frac{\text{cn}(2K'(m-\frac{1}{2})/n, k')}{\text{sn}(2K'(m-\frac{1}{2})/n, k')}\right]^2.$$

We also know that $M = \prod_{m=1}^{\lfloor \frac{1}{2}n \rfloor}(1-c_m)/(1-c'_m)$, and $1/\lambda = (\bar{\xi}/M)\prod_{m=1}^{\lfloor \frac{1}{2}n \rfloor}(1-c_m\bar{\xi}^2)/(1-c'_m\bar{\xi}^2)$ with $\bar{\xi} \equiv \text{sn}(K+iK'/n, k)$. We may use the addition formula to express the Jacobi elliptic functions of complex argument in terms of ones with purely real or imaginary arguments, so

$$\bar{\xi} = \text{sn}\left(K + \frac{iK'}{n}, k\right) = \frac{\text{sn}\,K\,\text{cn}\,\frac{iK'}{n}\,\text{dn}\,\frac{iK'}{n} + \text{sn}\,\frac{iK'}{n}\,\text{cn}\,K\,\text{dn}\,K}{1 - \left(k\,\text{sn}\,K\,\text{sn}\,\frac{iK'}{n}\right)^2}$$

$$= \frac{\text{cn}\,\frac{iK'}{n}\,\text{dn}\,\frac{iK'}{n}}{1 - \left(k\,\text{sn}\,\frac{iK'}{n}\right)^2} = \frac{\text{cn}\,\frac{iK'}{n}}{\text{dn}\,\frac{iK'}{n}}.$$

These may be converted to expressions involving only real arguments by the use of Jacobi's imaginary transformation (7),

$$\text{sn}\left(\frac{iK'}{n}, k\right) = \frac{i\,\text{sn}\left(\frac{K'}{n}, k'\right)}{\text{cn}\left(\frac{K'}{n}, k'\right)},$$

$$\text{cn}\left(\frac{iK'}{n}, k\right) = \sqrt{1 + \left[\frac{\text{sn}\left(\frac{K'}{n}, k'\right)}{\text{cn}\left(\frac{K'}{n}, k'\right)}\right]^2} = \frac{1}{\text{cn}\left(\frac{K'}{n}, k'\right)},$$

$$\text{dn}\left(\frac{iK'}{n}, k\right) = \sqrt{1 + \left[\frac{k\,\text{sn}\left(\frac{K'}{n}, k'\right)}{\text{cn}\left(\frac{K'}{n}, k'\right)}\right]^2} = \frac{\text{dn}\left(\frac{K'}{n}, k'\right)}{\text{cn}\left(\frac{K'}{n}, k'\right)},$$

giving the simple result $\bar{\xi} = 1/\text{dn}(K'/n, k')$.

Putting these results together we have

$$R(x) = Ax \prod_{m=1}^{\lfloor \frac{1}{2}n \rfloor} \frac{x^2 - a_m}{x^2 - a'_m}$$

with

$$a_m = \frac{k^2}{c_m} = -\left[k\frac{\text{sn}\left(\frac{2K'm}{n}, k'\right)}{\text{cn}\left(\frac{2K'm}{n}, k'\right)}\right]^2, \quad a'_m = \frac{k^2}{c'_m} = -\left[k\frac{\text{sn}\left(\frac{2K'(m-\frac{1}{2})}{n}, k'\right)}{\text{cn}\left(\frac{2K'(m-\frac{1}{2})}{n}, k'\right)}\right]^2,$$

$$A = \frac{2}{1+1/\lambda}\frac{1}{k}\prod_{m=1}^{\lfloor \frac{1}{2}n \rfloor}\frac{c_m}{c'_m}\left(\frac{1-c'_m}{1-c_m}\right), \quad \Delta = \frac{1-\lambda}{1+\lambda},$$

where Δ is the maximum error of the approximation.

Acknowledgements

I would like to thank Urs Wenger for helpful discussions, and Waseem Kamleh for correcting the quite unnecessarily obfuscated and previously incorrect bit manipulations in Fig. 1.

References

1. H. Neuberger, Phys. Lett. **B417**, 141 (1998), `hep-lat/9707022`.
2. H. Neuberger, Phys. Rev. Lett. **81**, 4060 (1998), `hep-lat/9806025`.
3. R. G. Edwards, U. M. Heller, and R. Narayanan, Nucl. Phys. **B540**, 457 (1999), `hep-lat/9807017`.
4. A. Boriçi, A. D. Kennedy, B. J. Pendleton, and U. Wenger, The overlap operator as a continued fraction, Nuclear Physics (Proceedings Supplements) Vol. B106–107, pp. 757–759, 2002, `hep-lat/0110070`, Proceedings of the XIXth International Symposium Lattice Field Theory, Berlin, Germany, 19–24 August 2001.
5. A. D. Kennedy, Approximation theory for matrices, Nuclear Physics B (Proceedings Supplements) Vol. 128C, pp. 107–116, 2004, `hep-lat/0402037`.
6. T. J. Rivlin, *An Introduction to the Approximation of Functions* (Dover, 1981).
7. Naum Il'ich Akhiezer, *Theory of Approximation* (Dover, 2004).
8. E. Y. Remez, *General Computational Methods of Chebyshev Approximation* (US Atomic Energy Commission, 1962).
9. E. W. Cheney, *Introduction to Approximation Theory*, 2nd ed. (American Mathematical Society, 2000).
10. E. I. Zolotarev, Zap. Imp. Akad. Nauk St. Petersburg **30**, 5 (1877).
11. Naum Il'ich Akhiezer, *Elements of the Theory of Elliptic Functions* Vol. 79 (AMS, 1990).
12. E. T. Whittaker and G. N. Watson, *A Course of Modern Analysis*, Fourth ed. (Cambridge University Press, 1927).
13. L. Carroll, *The Annotated Alice* (Wings Books, 1993), Edited by Martin Gardner.

The Overlap Dirac Operator as a Continued Fraction

Urs Wenger[1,2]

[1] Theoretical Physics, Oxford University, 1 Keble Road, Oxford OX1 3NP, UK
[2] NIC/DESY Zeuthen, Platanenallee 6, D–15738 Zeuthen, Germany
urs.wenger@desy.de

Summary. We use a continued fraction expansion of the sign–function in order to obtain a five dimensional formulation of the overlap lattice Dirac operator. Within this formulation the inverse of the overlap operator can be calculated by a single Krylov space method and nested conjugate gradient procedures are avoided. We point out that the five dimensional linear system can be made well conditioned using equivalence transformations on the continued fractions.

1 Introduction

Let us start with noting the overlap Dirac operator D describing chirally symmetric fermions on the lattice [1],

$$D = \frac{1}{2}\Big(1 + \gamma_5 \text{sign}\big(H(-m)\big)\Big), \qquad (1)$$

where $H(-m)$ is a hermitian Dirac operator with a negative mass parameter $-m$ of the order of the cut-off of the lattice theory, $m \sim O(1/a)$. A bare quark mass μ is most conveniently introduced as $D(\mu) = (1-\mu)D + \mu$. In order to calculate efficiently the sign–function in eq.(1) one can use a rational approximation $\text{sign}(x) \simeq R_{n,m}(x)$ where $R_{n,m}$ is a (nondegenerate and irreducible) rational function with algebraic polynomials of order n and m as numerator and denominator, respectively. Rational approximations usually converge much faster with their degree than polynomial approximations, but of course for our specific application it might still be much more expensive to apply the low degree denominator of the rational function than to apply a high order polynomial. However, noting that a rational function can be expanded as a partial fraction by matching poles and residues, i.e., $R_{n,m}(x) \sim x \sum_k c_k/(x^2 + q_k)$, one can use a multi–shift linear system solver where the convergence is governed by the smallest of the shifts q_k. The overall cost is therefore roughly equivalent to one inversion of H^2.

In order to do physics we need to compute inverses of the overlap operator $D(\mu)$ to obtain propagators, fermionic forces for Hybrid Monte Carlo, etc. If we consider the multi–shift linear system above, we realise that the inversion of $D(\mu)$ leads to a two–level nested linear system solution, which is rather cumbersome and forbidding.

It is well known how this can be avoided [2, 3]: by introducing a continued fraction representation of the rational approximation and auxiliary fields, the non–linear system $D(\mu)\psi = \chi$ can be unfolded into a set of systems linear in H. The auxiliary fields can be interpreted as fields living in a fictitious fifth dimension and in this way the nested 4D Krylov space problem reduces to finding a solution in a single 5D Krylov space. One can also regard the auxiliary fields as additional fermion flavours which, when integrated out, generate an effective Dirac operator equivalent to $D(\mu)$.

2 Matrix Representation of the Sign–function

Consider a rational approximation to the sign–function and expand it as a continued fraction[3],

$$R_{2n+1,2n}(x) = \alpha_0 x + \cfrac{\alpha_1}{x + \cfrac{\cdots}{\cdots + \cfrac{\alpha_{2n}}{x}}}. \qquad (2)$$

If we rewrite the linear system

$$\left(\alpha_0 x + \cfrac{\alpha_1}{x + \cfrac{\cdots}{\cdots + \frac{\alpha_{2n}}{x}}}\right)\psi = \chi \qquad (3)$$

using appropriate auxiliary fields $\phi_1, \phi_2, \ldots, \phi_{2n}$ we obtain the system

$$\begin{pmatrix} \alpha_0 x & \sqrt{\alpha_1} & & & & \\ \sqrt{\alpha_1} & -x & \sqrt{\alpha_2} & & & \\ & \sqrt{\alpha_2} & x & & & \\ & & & \ddots & & \\ & & & & -x & \sqrt{\alpha_{2n}} \\ & & & & \sqrt{\alpha_{2n}} & x \end{pmatrix} \begin{pmatrix} \psi \\ \phi_1 \\ \phi_2 \\ \vdots \\ \phi_{2n-1} \\ \phi_{2n} \end{pmatrix} = \begin{pmatrix} \chi \\ 0 \\ 0 \\ \vdots \\ 0 \\ 0 \end{pmatrix}. \qquad (4)$$

By performing explicitly a UDL decomposition it is easy to see that eq.(2) is just the Schur complement with respect to the $(2:2n+1, 2:2n+1)$ block of the matrix, and that eq.(4) reduces to eq.(3).

The rational function can also be mapped into a so–called simple continued fraction

$$R_{2n+1,2n}(x) = \beta_0 x + \cfrac{1}{\beta_1 x + \cfrac{\cdots}{\cdots + \cfrac{1}{\beta_{2n} x}}}$$

which leads to a different matrix

---

[3]Note that since polynomial approximations can also be expanded into continued fractions all our considerations apply to them as well.

$$\begin{pmatrix} \beta_0 x & 1 & & & & \\ 1 & -\beta_1 x & 1 & & & \\ & 1 & \beta_2 x & & & \\ & & & \ddots & & \\ & & & & -\beta_{2n-1}x & 1 \\ & & & & 1 & \beta_{2n}x \end{pmatrix},$$

and we find that the α_i's and β_i's are related through

$$\beta_0 = \alpha_0, \beta_1 = \frac{1}{\alpha_1}, \ldots, \beta_i = \frac{1}{\alpha_i \beta_{i-1}}, \ldots,$$

$$\alpha_0 = \beta_0, \alpha_1 = \frac{1}{\beta_1}, \ldots, \alpha_i = \frac{1}{\beta_{i-1}\beta_i}, \ldots.$$

In order to understand the relation between different continued fraction representations of the same rational function in detail we need to take a closer look at the properties of continued fractions (see e.g.[4, 5]).

3 Continued Fractions

A generic (truncated) continued fraction $\frac{A_n}{B_n}$ is conveniently written as

$$\frac{A_n}{B_n} = \beta_0 + \frac{\alpha_1|}{|\beta_1} + \frac{\alpha_2|}{|\beta_2} \ldots + \frac{\alpha_n}{\beta_n}.$$

Simple continued fractions have the property $\alpha_i = 1, i = 1, ..., n$ and one usually writes

$$\frac{A_n}{B_n} = [\beta_0; \beta_1, \beta_2, \ldots, \beta_n].$$

Continued fractions are widely used in many areas of physics and mathematics, in particular also in number theory. Finite continued fractions provide an alternative representation of rational numbers and form the basis of rational approximation theory. Infinite continued fractions on the other hand can be used to represent irrational numbers. Some numbers have beautiful continued fraction expansions while others have very mysterious ones. Let us quickly note a few examples for our amusement:

$$\phi = [1; 1, 1, 1, 1, \ldots],$$
$$\sqrt{2} = [1; 2, 2, 2, 2, \ldots],$$
$$e = [2; 1, 2, 1, 1, 4, 1, 1, 6, 1, 1, 8, 1, 1, 10, 1, \ldots],$$
$$\pi = [3; 7, 15, 1, 292, 1, 1, 1, 2, 1, 3, 1, \ldots],$$

where $\phi = \frac{1}{2}(1 + \sqrt{5})$ is the golden mean. It is interesting to note that there is no regular pattern known for π, and it is not known why this is so.

Evaluation of continued fractions can be done through forward or backward recurrence algorithms, and the former makes use of the intimate relation between continued fractions and the coupled two term relations

$$A_{-1} = 1, \; A_0 = \beta_0, \; B_{-1} = 0, \; B_0 = 1,$$
$$A_i = \beta_i A_{i-1} + \alpha_i A_{i-2}, \quad i = 1, 2, 3, \ldots,$$
$$B_i = \beta_i B_{i-1} + \alpha_i B_{i-2}, \quad i = 1, 2, 3, \ldots,$$

which can equivalently be written as an iterative matrix equation,

$$\begin{pmatrix} A_i \\ B_i \end{pmatrix} = \begin{pmatrix} A_{i-1} & A_{i-2} \\ B_{i-1} & B_{i-2} \end{pmatrix} \begin{pmatrix} \beta_i \\ \alpha_i \end{pmatrix}.$$

There is also an interesting connection between continued fractions, Moebius transformations and the corresponding unimodular matrices.

The natural arithmetic operation for continued fractions is inversion and the corresponding rule is particularly simple:

$$[\beta_0; \beta_1, \ldots]^{-1} = \begin{cases} [0; \beta_0, \beta_1, \ldots] & \text{if } \beta_0 \neq 0, \\ [\beta_1; \beta_2, \ldots] & \text{if } \beta_0 = 0. \end{cases}$$

It is also helpful to write down the rule for the multiplication of a continued fraction by a constant c,

$$c \cdot [\beta_0; \beta_1, \ldots, \beta_n] = [c \cdot \beta_0; \frac{\beta_1}{c}, c \cdot \beta_2, \frac{\beta_3}{c}, c \cdot \beta_4, \ldots].$$

Most important for our purpose, however, is the equivalence transformation of a continued fraction which is stated in the following theorem:

Theorem 1. *Two continued fractions $\beta_0 + \frac{\alpha_1|}{|\beta_1} + \ldots + \frac{\alpha_n}{\beta_n}$ and $\beta_0' + \frac{\alpha_1'|}{|\beta_1'} + \ldots + \frac{\alpha_n'}{\beta_n'}$ are equivalent iff there exists a sequence of non–zero constants c_n with $c_0 = 1$ such that*

$$\alpha_i' = c_i c_{i-1} \alpha_i, \quad i = 1, 2, 3, \ldots, n,$$
$$\beta_i' = c_i \beta_i, \quad i = 0, 1, 2, \ldots, n.$$

The theorem is easily seen to hold true by explicitly writing out the full continued fraction,

$$\beta_0 + \cfrac{\alpha_1}{\beta_1 + \cfrac{\alpha_2}{\beta_2 + \cfrac{\cdots}{\cdots + \cfrac{\alpha_n}{\beta_n}}}} = \beta_0 + \cfrac{c_1 \alpha_1}{c_1 \beta_1 + \cfrac{c_1 c_2 \alpha_2}{c_2 \beta_2 + \cfrac{\cdots}{\cdots + \cfrac{c_{n-1} c_n \alpha_n}{c_n \beta_n}}}}$$

and it is also clear that in terms of approximants we simply have $\frac{A_n}{B_n} = \frac{c_1 c_2 \ldots c_n A_n}{c_1 c_2 \ldots c_n B_n}$. So we find that a given rational function corresponds to an equivalence class of continued fractions and the class is parametrised by the (non–zero) coefficients c_i. While the equivalence transformation itself appears to be rather trivial, and indeed leaves the value of the continued fraction invariant, it affects the spectrum of the corresponding matrix representation [3].

In the analytic theory of continued fractions one considers continued fractions of the form

$$\beta_0(z) + \frac{\alpha_1(z)|}{|\beta_1(z)} + \frac{\alpha_2(z)|}{|\beta_2(z)} + \ldots,$$

i.e., the coefficients are functions of a complex variable z. So–called J–fractions are of the special form

$$[0; r_1 z + s_1, r_2 z + s_2, r_3 z + s_3, \ldots],$$

where r_i, s_i are complex numbers with $r_i \neq 0$. One can show that the n-th approximant of a J–fraction is an element of $R_{n-1,n}$. Conversely, let

$$P_{n-1}(z) = p_1 z^{n-1} + p_2 z^{n-2} + \ldots + p_n,$$
$$Q_n(z) = q_0 z^n + q_1 z^{n-1} + \ldots + q_n,$$

then we have

$$\frac{P_{n-1}(z)}{Q_n(z)} = [0; r_1 z + s_1, r_2 z + s_2, \ldots, r_n z + s_n]$$

where r_i, s_i are uniquely determined by p_i, q_i. We can now replace z by $-z$, apply an appropriate equivalence transformation and, using the fact that $P_{n-1}(z)/Q_n(z)$ is odd, we find $s_i = 0$ from uniqueness. Therefore we can always write

$$z \frac{P_{n-1}(z^2)}{Q_n(z^2)} = [0; k_1 z, k_2 z, k_3 z, \ldots, k_m z],$$

where $m = 2n - 1$ for $Q_n(0) = 0$, or $m = 2n$ for $Q_n(0) \neq 0$.

4 Application to the Overlap Operator

We are now in a position to apply our knowledge to find the solution to the equation $\frac{2}{1-\mu}\gamma_5 D(\mu)\psi = \chi$. Collecting the results from the previous two sections we obtain an equivalence class of five dimensional linear systems

$$\begin{pmatrix} A\gamma_5 + k_0 H & c_1 & & & \\ c_1 & -c_1^2 k_1 H & c_1 c_2 & & \\ & c_1 c_2 & c_2^2 k_2 H & & \\ & & & \ddots & \\ & & & -c_{2n-1}^2 k_{2n-1} H & c_{2n-1} c_{2n} \\ & & & c_{2n-1} c_{2n} & c_{2n}^2 k_{2n} H \end{pmatrix} \begin{pmatrix} \psi \\ \phi_1 \\ \phi_2 \\ \vdots \\ \phi_{2n-1} \\ \phi_{2n} \end{pmatrix} = \begin{pmatrix} \chi \\ 0 \\ 0 \\ \vdots \\ 0 \\ 0 \end{pmatrix}$$

where the k_i's and n are uniquely determined by the given rational approximation to the sign–function, the c_i's parametrise the corresponding equivalence class and $A = \frac{1+\mu}{1-\mu}$. It is now crucial to see how the spectrum of the five dimensional matrix depends on the free parameters c_i as we already emphasised in the last section. While for a generic set of parameters the five dimensional system is usually ill–defined, the condition number can be kept under control with a clever choice of c_i's [3, 6] enabling one to optimise the matrix, e.g., for fast inversions. As was pointed out in [3] the equivalence transformations can be understood as a block Jacobi preconditioning without any computational overhead. The particularly simple structure of the five dimensional operator allows improvements in various directions: one can easily change and optimise the hermitian overlap kernel H or apply various well known preconditioning techniques such as an even–odd or ILU decomposition [6].

It is instructive to see that the first auxiliary field disentangles γ_5 from the sign–function, i.e., $(A \cdot \gamma_5 + \text{sign}(H))\psi = \chi$ maps into

$$\begin{pmatrix} A \cdot \gamma_5 & c \\ c & -c^2 \text{sign}(H) \end{pmatrix} \begin{pmatrix} \psi \\ \phi \end{pmatrix} = \begin{pmatrix} \chi \\ 0 \end{pmatrix}$$

where we used sign = sign^{-1}. Additional fields are then only used to generate the sign–function (or its approximation, respectively). We now have two systems which can be easily solved,

$$\psi = \frac{1}{A}\gamma_5(\chi - c\phi), \quad \phi = \frac{1}{c}\text{sign}(H)\psi,$$

yielding recursion relations for ψ and ϕ,

$$\psi^{(i+1)} = \frac{1}{A}\gamma_5(\chi - \text{sign}(H)\psi^{(i)}), \quad \phi^{(i+1)} = \frac{1}{A}\text{sign}(H)\gamma_5(\frac{1}{c}\chi - \phi^{(i)}).$$

Equivalently, there are recursion relations for the residuals and one can show that

$$|r^{(i)}_{\psi,\phi}| = \left(\frac{1-\mu}{1+\mu}\right)^i |r^{(0)}_{\psi,\phi}|.$$

Of course one can use a similar recursive scheme for the case where $\text{sign}(H)$ is expressed as a continued fraction and one has several auxiliary fields.

Projection of eigenvectors of H close to 0 is a valuable tool to improve approximations to the overlap operator and here it is straightforward to implement. However, we wish to point out that in this formulation it might not be necessary at all. Consider the linear system in the lower right corner,

$$c_{2n-1}c_{2n}\phi_{2n-1} - c^2_{2n}k_{2n}H\phi_{2n} = 0. \tag{5}$$

For a typical rational approximation to the sign–function we have $k_{2n} \gg 1$ and with the choice $c_{2n} \simeq 1/\sqrt{k_{2n+1}} \ll 1$. So it turns out that the system in eq.(5) is essentially equivalent to finding eigenvectors of H close to zero, i.e., the two Krylov spaces possibly have a large overlap. Indeed a truncation in the fifth dimension, i.e., of a given continued fraction, changes the approximation only in the neighbourhood around $H = 0$ and therefore provides a natural truncation scheme for approximations of fixed accuracy. It also opens up the possibility of applying successively better approximations to the overlap operator which are ultra–local in five dimensions.

5 Summary and Outlook

We have shown how to use a continued fraction expansion of the sign–function in order to obtain a five dimensional formulation of the overlap lattice Dirac operator. We have pointed out that the operator can be made well conditioned using equivalence transformations on the continued fractions. It is now important to investigate in detail strategies to exploit this freedom in practical applications and such a study is under way [6]. If successful, and first results indicate that this is indeed the case, the formulation would provide a valuable alternative for the simulation of dynamical chiral fermions.

6 Acknowledgments

I would like to thank the organisers of the workshop for the invitation and for creating a stimulating atmosphere. I also thank Tony Kennedy for discussions and comments. This research was supported by a PPARC SPG fellowship.

References

1. H. Neuberger, Phys. Lett. **B417**, 141–144, (1998), hep-lat/9707022.
2. H. Neuberger, Phys. Rev. **D60** 065006 (1999), hep-lat/9901003.
3. A. Borici, A. D. Kennedy, B. J. Pendleton and U. Wenger, Nucl. Phys. Proc. Suppl. **106**, 757–759, (2002), hep-lat/0110070.
4. H. Wall, Analytic theory of continued fractions (Chelsea Publishing company 1948).
5. W. Jones and W. Thron, Continued fractions: analytic theory and applications, Vol. 11 (Addison-Wesley, 1980).
6. U. Wenger, in progress.

Index

action, asqtad   77
action, fat7   77
action, HYP   77
action, Symanzik improved   84
Aldor programming language   127
approximate matrix-vector products   135, 138
Arnoldi process   160
Arnoldi relation   124

baryon sector   103
Bi-CGstab   134
BiCG   125
BiDLanczos   125

Cauchy integral theorem   20
central order statistics   60
CG   9, 123, 139, 159, 165
CGNE   162
CGR   139
chaos   94
Chebyshev   169, 183
chiral extrapolation   118
chiral perturbation theory   113
chiral projection approach   159
chiral symmetry   6, 34, 42
compiler optimisation   130
complete elliptic integral   170–187
continued fraction representation   192
continued fraction, simple   192
cross-component optimisation   130
CS-decomposition   155

detailed balance   45, 106

determinant, Gaussian integral representation   28, 57
Dirac-Clifford algebra   26
Dirac/Weyl fields   3
DLanczos   125
doubling problem   76
dynamical simulation   96, 107, 113, 154, 159

effective field theory   113
effective masses method   146
ergodicity   44, 92, 106
exact algorithms   49
expression templates   126

fat link   52
fermion determinant   27, 28, 61, 106
fermions, chiral   34, 196
fermions, domain wall   10, 34, 36
fermions, dynamical   91, 196
fermions, overlap   34, 36, 153
fermions, staggered   42, 78, 144
fermions, truncated overlap   36
fermions, Wilson   144
finite density   101
finite range regularisation   113
FOM   123–125
framework categories   128

Galerkin extraction   134, 136
Galerkin method   125, 126
gauge action   27, 63, 106
gauge fields   4
gauge invariance   4, 52

Gauss   169, 183–187
GCR   139
Ginsparg-Wilson relation   8
GMERR   125
GMRES   123–125, 139, 160
Grassmann field   3, 26, 45

hadronic correlations   143
heatbath   68
Hybrid Monte Carlo algorithm   49, 91, 92, 191

imaginary chemical potential   102
importance sampling   43
inner-outer paradigm   153
instability   93–96
IOM(k)   123

Jacobi elliptic functions   180–188
Jacobi iteration   37

Kentucky Noisy Monte Carlo Algroithm   107
Kogut-Susskind operator   62
Krylov space   29, 124, 129, 134
Krylov subspace method   50, 96, 134, 153
Krylov subspace method, inexact   135, 137

Lanczos algorithm   29, 31
Lanczos process, two-sided   124, 129
Landau gauge   83, 87
Landau mean link scheme   86
large-N limit   67
lazy matrix   129
least squares, nonlinear   145
linear algebra categories   128
linear partial fraction form   16
linear prediction method   147
Liouville   170–176

Markov Chain Monte Carlo   44
Markov process   43
matrix functions   15
matrix logarithm   16, 29, 58
matrix power   20
matrix sign function   9, 153, 191
Metropolis   68
Metropolis algorithm   45, 63

Metropolis test   92, 109
minimal residual extraction   134, 136
minimum error method   125
minimum residual method   125, 126
MINRES   125, 160–165
molecular dynamics   47
molecular dynamics integrator   91, 94
Monte Carlo   67
Monte Carlo algorithm, noisy   105, 107
Monte Carlo method   28, 84, 143
Monte Carlo simulation   43, 87
multigrid algorithm   37
multiple shifts   9

nested iteration   37, 191
NMR spectroscopy   144
noisy estimation   28, 57, 59
nucleon mass expansion   114

order statistics   59
ORTHORES   139
overlap Dirac operator   6, 191
overlap operator   34, 36, 42
overlap operator, massive   154, 167
overlap problem   102, 104
overrelaxation   68

Padé approximation   17, 30, 107
Padé-Z_2 estimates   105
partial fraction expansion   30, 107, 191
path integral   25, 27, 41, 63
pion spectrum   78
plaquette   63
preconditioner   36
preconditioning   51
preconditioning, block Jacobi   195
pseudofermions   28, 46
pseudospectra   22

QMR   123, 125
quadrature for matrix functions   17
quadrature, adaptive   18
quadrature, Gaussian   17
quadrature, trapezium rule   21
quark mass, bare   26, 42
quenched approximation   27
quenched QCD   95
quenched simulation   104

R-algorithm   49
rational approximation   16, 36, 191

relaxation strategy   140
renormalised chiral expansion   116
residual gap   135, 139
resolution   5
Ritz value   30

Schwinger model   62
shifted linear systems   31
sign problem   101
signum function   169, 171, 183
simple harmonic oscillator   93
staggered fermion matrix   42
staggered fermions, improved   75
state space methods   149
stochastic estimate   43
stochastic estimator   57, 106
stochastic estimator, biased   59, 62
stochastic estimator, unbiased   61, 62
SU(N)   67
SUMR   160–165
SYMMLQ   123–126
symplectic integrators   93

tadpole improvement   83
taste-changing interactions   75, 77

total least squares   149
twisted boundary conditions   87
two-level algorithm   37
two-loop formula   84
two-sweep approach   165
type system (Aldor language)   127

variable projection functional   146

Weierstrass   170
Weierstrass elliptic function   172–181
Weierstrass Q function   172
Weierstrass sigma function   174, 175
Weierstrass zeta function   174–181
Weyl operators   4
Wilson fermion matrix   42
Wilson-Dirac operator   6, 26, 28, 62, 166
Wilson-Dirac operator, hermitian form   7, 26, 153, 166

Yang-Mills   67

Zolotarev   169–189
Zolotarev approximation   9, 36, 154

Editorial Policy

§1. Volumes in the following three categories will be published in LNCSE:

i) Research monographs
ii) Lecture and seminar notes
iii) Conference proceedings

Those considering a book which might be suitable for the series are strongly advised to contact the publisher or the series editors at an early stage.

§2. Categories i) and ii). These categories will be emphasized by Lecture Notes in Computational Science and Engineering. **Submissions by interdisciplinary teams of authors are encouraged.** The goal is to report new developments – quickly, informally, and in a way that will make them accessible to non-specialists. In the evaluation of submissions timeliness of the work is an important criterion. Texts should be well-rounded, well-written and reasonably self-contained. In most cases the work will contain results of others as well as those of the author(s). In each case the author(s) should provide sufficient motivation, examples, and applications. In this respect, Ph.D. theses will usually be deemed unsuitable for the Lecture Notes series. Proposals for volumes in these categories should be submitted either to one of the series editors or to Springer-Verlag, Heidelberg, and will be refereed. A provisional judgment on the acceptability of a project can be based on partial information about the work: a detailed outline describing the contents of each chapter, the estimated length, a bibliography, and one or two sample chapters – or a first draft. A final decision whether to accept will rest on an evaluation of the completed work which should include

– at least 100 pages of text;
– a table of contents;
– an informative introduction perhaps with some historical remarks which should be accessible to readers unfamiliar with the topic treated;
– a subject index.

§3. Category iii). Conference proceedings will be considered for publication provided that they are both of exceptional interest and devoted to a single topic. One (or more) expert participants will act as the scientific editor(s) of the volume. They select the papers which are suitable for inclusion and have them individually refereed as for a journal. Papers not closely related to the central topic are to be excluded. Organizers should contact Lecture Notes in Computational Science and Engineering at the planning stage.

In exceptional cases some other multi-author-volumes may be considered in this category.

§4. Format. Only works in English are considered. They should be submitted in camera-ready form according to Springer-Verlag's specifications.
Electronic material can be included if appropriate. Please contact the publisher.
Technical instructions and/or TeX macros are available via
http://www.springeronline.com/sgw/cda/frontpage/0,10735,5-111-2-71391-0,00.html
The macros can also be sent on request.

General Remarks

Lecture Notes are printed by photo-offset from the master-copy delivered in camera-ready form by the authors. For this purpose Springer-Verlag provides technical instructions for the preparation of manuscripts. See also *Editorial Policy*.

Careful preparation of manuscripts will help keep production time short and ensure a satisfactory appearance of the finished book.

The following terms and conditions hold:

Categories i), ii), and iii):
Authors receive 50 free copies of their book. No royalty is paid. Commitment to publish is made by letter of intent rather than by signing a formal contract. Springer-Verlag secures the copyright for each volume.

For conference proceedings, editors receive a total of 50 free copies of their volume for distribution to the contributing authors.

All categories:
Authors are entitled to purchase further copies of their book and other Springer mathematics books for their personal use, at a discount of 33,3 % directly from Springer-Verlag.

Addresses:

Timothy J. Barth
NASA Ames Research Center
NAS Division
Moffett Field, CA 94035, USA
e-mail: barth@nas.nasa.gov

Michael Griebel
Institut für Angewandte Mathematik
der Universität Bonn
Wegelerstr. 6
53115 Bonn, Germany
e-mail: griebel@ins.uni-bonn.de

David E. Keyes
Department of Applied Physics
and Applied Mathematics
Columbia University
200 S. W. Mudd Building
500 W. 120th Street
New York, NY 10027, USA
e-mail: david.keyes@columbia.edu

Risto M. Nieminen
Laboratory of Physics
Helsinki University of Technology
02150 Espoo, Finland
e-mail: rni@fyslab.hut.fi

Dirk Roose
Department of Computer Science
Katholieke Universiteit Leuven
Celestijnenlaan 200A
3001 Leuven-Heverlee, Belgium
e-mail: dirk.roose@cs.kuleuven.ac.be

Tamar Schlick
Department of Chemistry
Courant Institute of Mathematical
Sciences
New York University
and Howard Hughes Medical Institute
251 Mercer Street
New York, NY 10012, USA
e-mail: schlick@nyu.edu

Springer-Verlag, Mathematics Editorial IV
Tiergartenstrasse 17
D-69121 Heidelberg, Germany
Tel.: *49 (6221) 487-8185
Fax: *49 (6221) 487-8355
e-mail: Martin.Peters@springer-sbm.com

Lecture Notes
in Computational Science
and Engineering

Vol. 1 D. Funaro, *Spectral Elements for Transport-Dominated Equations*. 1997. X, 211 pp. Softcover. ISBN 3-540-62649-2

Vol. 2 H. P. Langtangen, *Computational Partial Differential Equations*. Numerical Methods and Diffpack Programming. 1999. XXIII, 682 pp. Hardcover. ISBN 3-540-65274-4

Vol. 3 W. Hackbusch, G. Wittum (eds.), *Multigrid Methods V*. Proceedings of the Fifth European Multigrid Conference held in Stuttgart, Germany, October 1-4, 1996. 1998. VIII, 334 pp. Softcover. ISBN 3-540-63133-X

Vol. 4 P. Deuflhard, J. Hermans, B. Leimkuhler, A. E. Mark, S. Reich, R. D. Skeel (eds.), *Computational Molecular Dynamics: Challenges, Methods, Ideas*. Proceedings of the 2nd International Symposium on Algorithms for Macromolecular Modelling, Berlin, May 21-24, 1997. 1998. XI, 489 pp. Softcover. ISBN 3-540-63242-5

Vol. 5 D. Kröner, M. Ohlberger, C. Rohde (eds.), *An Introduction to Recent Developments in Theory and Numerics for Conservation Laws*. Proceedings of the International School on Theory and Numerics for Conservation Laws, Freiburg / Littenweiler, October 20-24, 1997. 1998. VII, 285 pp. Softcover. ISBN 3-540-65081-4

Vol. 6 S. Turek, *Efficient Solvers for Incompressible Flow Problems*. An Algorithmic and Computational Approach. 1999. XVII, 352 pp, with CD-ROM. Hardcover. ISBN 3-540-65433-X

Vol. 7 R. von Schwerin, *Multi Body System SIMulation*. Numerical Methods, Algorithms, and Software. 1999. XX, 338 pp. Softcover. ISBN 3-540-65662-6

Vol. 8 H.-J. Bungartz, F. Durst, C. Zenger (eds.), *High Performance Scientific and Engineering Computing*. Proceedings of the International FORTWIHR Conference on HPSEC, Munich, March 16-18, 1998. 1999. X, 471 pp. Softcover. 3-540-65730-4

Vol. 9 T. J. Barth, H. Deconinck (eds.), *High-Order Methods for Computational Physics*. 1999. VII, 582 pp. Hardcover. 3-540-65893-9

Vol. 10 H. P. Langtangen, A. M. Bruaset, E. Quak (eds.), *Advances in Software Tools for Scientific Computing*. 2000. X, 357 pp. Softcover. 3-540-66557-9

Vol. 11 B. Cockburn, G. E. Karniadakis, C.-W. Shu (eds.), *Discontinuous Galerkin Methods*. Theory, Computation and Applications. 2000. XI, 470 pp. Hardcover. 3-540-66787-3

Vol. 12 U. van Rienen, *Numerical Methods in Computational Electrodynamics*. Linear Systems in Practical Applications. 2000. XIII, 375 pp. Softcover. 3-540-67629-5

Vol. 13 B. Engquist, L. Johnsson, M. Hammill, F. Short (eds.), *Simulation and Visualization on the Grid*. Parallelldatorcentrum Seventh Annual Conference, Stockholm, December 1999, Proceedings. 2000. XIII, 301 pp. Softcover. 3-540-67264-8

Vol. 14 E. Dick, K. Riemslagh, J. Vierendeels (eds.), *Multigrid Methods VI*. Proceedings of the Sixth European Multigrid Conference Held in Gent, Belgium, September 27-30, 1999. 2000. IX, 293 pp. Softcover. 3-540-67157-9

Vol. 15 A. Frommer, T. Lippert, B. Medeke, K. Schilling (eds.), *Numerical Challenges in Lattice Quantum Chromodynamics*. Joint Interdisciplinary Workshop of John von Neumann Institute for Computing, Jülich and Institute of Applied Computer Science, Wuppertal University, August 1999. 2000. VIII, 184 pp. Softcover. 3-540-67732-1

Vol. 16 J. Lang, *Adaptive Multilevel Solution of Nonlinear Parabolic PDE Systems*. Theory, Algorithm, and Applications. 2001. XII, 157 pp. Softcover. 3-540-67900-6

Vol. 17 B. I. Wohlmuth, *Discretization Methods and Iterative Solvers Based on Domain Decomposition*. 2001. X, 197 pp. Softcover. 3-540-41083-X

Vol. 18 U. van Rienen, M. Günther, D. Hecht (eds.), *Scientific Computing in Electrical Engineering*. Proceedings of the 3rd International Workshop, August 20-23, 2000, Warnemünde, Germany. 2001. XII, 428 pp. Softcover. 3-540-42173-4

Vol. 19 I. Babuška, P. G. Ciarlet, T. Miyoshi (eds.), *Mathematical Modeling and Numerical Simulation in Continuum Mechanics*. Proceedings of the International Symposium on Mathematical Modeling and Numerical Simulation in Continuum Mechanics, September 29 - October 3, 2000, Yamaguchi, Japan. 2002. VIII, 301 pp. Softcover. 3-540-42399-0

Vol. 20 T. J. Barth, T. Chan, R. Haimes (eds.), *Multiscale and Multiresolution Methods*. Theory and Applications. 2002. X, 389 pp. Softcover. 3-540-42420-2

Vol. 21 M. Breuer, F. Durst, C. Zenger (eds.), *High Performance Scientific and Engineering Computing*. Proceedings of the 3rd International FORTWIHR Conference on HPSEC, Erlangen, March 12-14, 2001. 2002. XIII, 408 pp. Softcover. 3-540-42946-8

Vol. 22 K. Urban, *Wavelets in Numerical Simulation*. Problem Adapted Construction and Applications. 2002. XV, 181 pp. Softcover. 3-540-43055-5

Vol. 23 L. F. Pavarino, A. Toselli (eds.), *Recent Developments in Domain Decomposition Methods*. 2002. XII, 243 pp. Softcover. 3-540-43413-5

Vol. 24 T. Schlick, H. H. Gan (eds.), *Computational Methods for Macromolecules: Challenges and Applications*. Proceedings of the 3rd International Workshop on Algorithms for Macromolecular Modeling, New York, October 12-14, 2000. 2002. IX, 504 pp. Softcover. 3-540-43756-8

Vol. 25 T. J. Barth, H. Deconinck (eds.), *Error Estimation and Adaptive Discretization Methods in Computational Fluid Dynamics*. 2003. VII, 344 pp. Hardcover. 3-540-43758-4

Vol. 26 M. Griebel, M. A. Schweitzer (eds.), *Meshfree Methods for Partial Differential Equations*. 2003. IX, 466 pp. Softcover. 3-540-43891-2

Vol. 27 S. Müller, *Adaptive Multiscale Schemes for Conservation Laws*. 2003. XIV, 181 pp. Softcover. 3-540-44325-8

Vol. 28 C. Carstensen, S. Funken, W. Hackbusch, R. H. W. Hoppe, P. Monk (eds.), *Computational Electromagnetics*. Proceedings of the GAMM Workshop on "Computational Electromagnetics", Kiel, Germany, January 26-28, 2001. 2003. X, 209 pp. Softcover. 3-540-44392-4

Vol. 29 M. A. Schweitzer, *A Parallel Multilevel Partition of Unity Method for Elliptic Partial Differential Equations*. 2003. V, 194 pp. Softcover. 3-540-00351-7

Vol. 30 T. Biegler, O. Ghattas, M. Heinkenschloss, B. van Bloemen Waanders (eds.), *Large-Scale PDE-Constrained Optimization*. 2003. VI, 349 pp. Softcover. 3-540-05045-0

Vol. 31 M. Ainsworth, P. Davies, D. Duncan, P. Martin, B. Rynne (eds.), *Topics in Computational Wave Propagation*. Direct and Inverse Problems. 2003. VIII, 399 pp. Softcover. 3-540-00744-X

Vol. 32 H. Emmerich, B. Nestler, M. Schreckenberg (eds.), *Interface and Transport Dynamics*. Computational Modelling. 2003. XV, 432 pp. Hardcover. 3-540-40367-1

Vol. 33 H. P. Langtangen, A. Tveito (eds.), *Advanced Topics in Computational Partial Differential Equations*. Numerical Methods and Diffpack Programming. 2003. XIX, 658 pp. Softcover. 3-540-01438-1

Vol. 34 V. John, *Large Eddy Simulation of Turbulent Incompressible Flows*. Analytical and Numerical Results for a Class of LES Models. 2004. XII, 261 pp. Softcover. 3-540-40643-3

Vol. 35 E. Bänsch (ed.), *Challenges in Scientific Computing - CISC 2002*. Proceedings of the Conference *Challenges in Scientific Computing*, Berlin, October 2-5, 2002. 2003. VIII, 287 pp. Hardcover. 3-540-40887-8

Vol. 36 B. N. Khoromskij, G. Wittum, *Numerical Solution of Elliptic Differential Equations by Reduction to the Interface*. 2004. XI, 293 pp. Softcover. 3-540-20406-7

Vol. 37 A. Iske, *Multiresolution Methods in Scattered Data Modelling*. 2004. XII, 182 pp. Softcover. 3-540-20479-2

Vol. 38 S.-I. Niculescu, K. Gu (eds.), *Advances in Time-Delay Systems*. 2004. XIV, 446 pp. Softcover. 3-540-20890-9

Vol. 39 S. Attinger, P. Koumoutsakos (eds.), *Multiscale Modelling and Simulation*. 2004. VIII, 277 pp. Softcover. 3-540-21180-2

Vol. 40 R. Kornhuber, R. Hoppe, J. Périaux, O. Pironneau, O. Wildlund, J. Xu (eds.), *Domain Decomposition Methods in Science and Engineering*. 2005. XVIII, 690 pp. Softcover. 3-540-22523-4

Vol. 41   T. Plewa, T. Linde, V.G. Weirs (eds.), *Adaptive Mesh Refinement – Theory and Applications.* 2005. XIV, 552 pp. Softcover. 3-540-21147-0

Vol. 42   A. Schmidt, K.G. Siebert, *Design of Adaptive Finite Element Software.* The Finite Element Toolbox ALBERTA. 2005. XII, 322 pp. Hardcover. 3-540-22842-X

Vol. 43   M. Griebel, M.A. Schweitzer (eds.), *Meshfree Methods for Partial Differential Equations II.* 2005. XIII, 303 pp. Softcover. 3-540-23026-2

Vol. 44   B. Engquist, P. Lötstedt, O. Runborg (eds.), *Multiscale Methods in Science and Engineering.* 2005. XII, 291 pp. Softcover. 3-540-25335-1

Vol. 45   P. Benner, V. Mehrmann, D.C. Sorensen (eds.), *Dimension Reduction of Large-Scale Systems.* 2005. XII, 402 pp. Softcover. 3-540-24545-6

Vol. 46   D. Kressner (ed.), *Numerical Methods for General and Structured Eigenvalue Problems.* 2005. XIV, 258 pp. Softcover. 3-540-24546-4

Vol. 47   A. Boriçi, A. Frommer, B. Joó, A. Kennedy, B. Pendleton (eds.), *QCD and Numerical Analysis III.* 2005. XIII, 201 pp. Softcover. 3-540-21257-4

For further information on these books please have a look at our mathematics catalogue at the following URL: www.springeronline.com/series/3527

Texts in Computational Science and Engineering

Vol. 1   H. P. Langtangen, *Computational Partial Differential Equations.* Numerical Methods and Diffpack Programming. 2nd Edition 2003. XXVI, 855 pp. Hardcover. ISBN 3-540-43416-X

Vol. 2   A. Quarteroni, F. Saleri, *Scientific Computing with MATLAB.* 2003. IX, 257 pp. Hardcover. ISBN 3-540-44363-0

Vol. 3   H. P. Langtangen, *Python Scripting for Computational Science.* 2004. XXII, 724 pp. Hardcover. ISBN 3-540-43508-5

For further information on these books please have a look at our mathematics catalogue at the following URL: www.springeronline.com/series/5151

Printing and Binding: Strauss GmbH, Mörlenbach